Water Transfers in the West

Efficiency, Equity, and the Environment

Committee on Western Water Management
Water Science and Technology Board
Commission on Engineering and Technical Systems

with the assistance of the
Board on Agriculture

National Research Council

NATIONAL ACADEMY PRESS
Washington, D.C. 1992

NATIONAL ACADEMY PRESS • 2101 Constitution Ave., N.W. • Washington, D.C. 20418

NOTICE: The project that is the subject of this report was approved by the Governing Board of the National Research Council, whose members are drawn from the councils of the National Academy of Sciences, the National Academy of Engineering, and the Institute of Medicine. The members of the committee responsible for the report were chosen for their special competences and with regard for appropriate balance.

This report has been reviewed by a group other than the authors according to procedures approved by a Report Review Committee consisting of members of the National Academy of Sciences, the National Academy of Engineering, and the Institute of Medicine.

Support for this project was provided by the Metropolitan Water District of Southern California Agreement No. 2103, East Bay Municipal Utility District, Bureau of Reclamation Contract No. 9-FG-81-15550, U.S. Environmental Protection Agency Contract No. 815848-01-O/R, and Ford Foundation Grant No. 890-0807.

Library of Congress Cataloging-in-Publication Data

Water transfers in the West: efficiency, equity, and the environment / Committee on Western Water Management, Water Science and Technology Board, Commission on Engineering and Technical Systems, Board on Agriculture, National Research Council.

p. cm.

Includes bibliographical references and index.

ISBN 0-309-04528-2

1. Water transfer—West (U.S.) 2. Water transfer—Law and legislation—West (U.S.) I. National Research Council (U.S.). Water Science and Technology Board. Committee on Western Water Management.

HD1695.A17W39 1992

333.91'00978—dc20 92-5740

CIP

This book is printed with soy ink on acid-free recycled stock.

Printed in the United States of America

Original cover art by Sally Groom, Washington, D.C.

COMMITTEE ON WESTERN WATER MANAGEMENT

A. DAN TARLOCK, *Chair,* Illinois Institute of Technology, Chicago-Kent College of Law
D. CRAIG BELL, Western States Water Council, Midvale, Utah
BONNIE COLBY, University of Arizona, Tucson
LEO M. EISEL, Wright Water Engineers, Denver, Colorado
DAVID H. GETCHES, University of Colorado School of Law, Boulder
THOMAS J. GRAFF, Environmental Defense Fund, Oakland, California
FRANK GREGG, University of Arizona, Tucson
R. KEITH HIGGINSON, Department of Water Resources, Boise, Idaho
MARVIN E. JENSEN, Colorado State University, Fort Collins
DUNCAN T. PATTEN, Arizona State University, Tempe
CLAIR B. STALNAKER, Fish and Wildlife Service, Fort Collins, Colorado
LUIS S. TORRES, Southwest Research Information Center, Espanola, New Mexico
RICHARD TRUDELL, American Indian Lawyer Training Program, Inc., Oakland, California
HENRY J. VAUX, JR., University of California, Riverside
SUSAN WILLIAMS, Gover, Stetson, Williams & West, Albuquerque, New Mexico

Liaison to the Board on Agriculture

JAN VAN SCHILFGAARDE, Agricultural Research Service, Beltsville, Maryland

Staff

CHRIS ELFRING, Study Director
ANITA A. HALL, Project Secretary
WYETHA B. TURNEY, Word Processor
ROSEANNE PRICE, Editorial Consultant
FLORENCE POILLON, Editorial Consultant

WATER SCIENCE AND TECHNOLOGY BOARD

Staff

STEPHEN D. PARKER, Director
SARAH CONNICK, Staff Officer
SHEILA D. DAVID, Senior Staff Officer
CHRIS ELFRING, Senior Staff Officer
JACQUELINE MACDONALD, Research Associate
JEANNE AQUILINO, Administrative Specialist
PATRICIA CICERO, Secretary
ANITA A. HALL, Administrative Secretary
JOYCE A. SPARROW, Secretary

The National Academy of Sciences is a private, nonprofit, self-perpetuating society of distinguished scholars engaged in scientific and engineering research, dedicated to the furtherance of science and technology and to their use for the general welfare. Upon the authority of the charter granted to it by the Congress in 1863, the Academy has a mandate that requires it to advise the federal government on scientific and technical matters. Dr. Frank Press is president of the National Academy of Sciences.

The National Academy of Engineering was established in 1964, under the charter of the National Academy of Sciences, as a parallel organization of outstanding engineers. It is autonomous in its administration and in the selection of its members, sharing with the National Academy of Sciences the responsibility for advising the federal government. The National Academy of Engineering also sponsors engineering programs aimed at meeting national needs, encourages education and research, and recognizes the superior achievements of engineers. Dr. Robert M. White is president of the National Academy of Engineering.

The Institute of Medicine was established in 1970 by the National Academy of Sciences to secure the services of eminent members of appropriate professions in the examination of policy matters pertaining to the health of the public. The Institute acts under the responsibility given to the National Academy of Sciences by its congressional charter to be an adviser to the federal government and, upon its own initiative, to identify issues of medical care, research, and education. Dr. Kenneth I. Shine is president of the Institute of Medicine.

The National Research Council was organized by the National Academy of Sciences in 1916 to associate the broad community of science and technology with the Academy's purposes of furthering knowledge and advising the federal government. Functioning in accordance with general policies determined by the Academy, the Council has become the principal operating agency of both the National Academy of Sciences and the National Academy of Engineering in providing services to the government, the public, and the scientific and engineering communities. The Council is administered jointly by both Academies and the Institute of Medicine. Dr. Frank Press and Dr. Robert M. White are chairman and vice chairman, respectively, of the National Research Council.

Preface

As yet another prolonged drought grips many parts of the arid West, we are once again reminded of the central importance of water. Water is a resource in great demand: beyond the needs of irrigated agriculture—long the biggest water user in the West—we now must ensure water supplies to support urban growth and development, traditional minority cultures, environmental needs, and recreation. We must become more careful in our planning and management so all these many needs can be met equitably and efficiently.

More and more, water transfers are being considered, and used, as a major component of water resource management. As defined in this report, water transfers involve a change in point of diversion, a change in type of use, or a change in location of use. They are a mechanism with great potential to help allocate the West's limited water. But, as is the case with all tools, water transfers can have various effects—sometimes beneficial, sometimes adverse; sometimes intended, sometimes inadvertent.

The Committee on Western Water Management, of the Water Science and Technology Board, has focused its attention on those effects. The committee's purpose was to look at water transfers that have happened or are being discussed and to develop an improved understanding of the nature, scope, and consequences of water transfers, with a special focus on third party effects and environmental quality. Although much of this report discusses these issues in broad terms, the committee also conducted seven case studies to provide examples that show the diversity of participants and impacts.

Water transfers have occurred in the western United States since the initiation of the prior appropriation doctrine. Many have been small scale and relatively routine, but others have been large and controversial. Some of these transfers and proposed transfers have provided the spark for significant controversy in the West. Examples include the Owens Valley transfer in California (dramatically described in the movie *Chinatown*), transmountain diversions from the Colorado River basin to Denver, and agriculture-to-municipal transfers in Arizona.

It is time to move beyond discussions of whether water transfers in general are good or bad per se. Transfers are a management option that western water managers have historically used along with structural measures to meet demand, and all indications are that transfers will be used with increasing frequency because of cost advantages and flexibility. We need to focus attention, then, on the consequences of such transfers and to think hard about how to encourage the benefits they can provide while minimizing the problems.

Unlike many other commodity transfers, water transfers can have significant positive and negative effects on "third parties"—people, groups, and values beyond the buyers and sellers involved. The law has afforded protection to some of these third parties—such as existing water rights holders not directly involved in the transfer—but not to all. Third party effects include impacts on Indian and other minority populations, effects on regional economic systems, and environmental effects. This report makes suggestions to help increase the usefulness of water transfers in solving water supply problems in the West while ensuring that third party effects are fairly considered and mitigated. Conflicts over water seem an inevitable part of life in an arid environment, but we can take steps to ease these controversies. The committee hopes this report provides some guidance to the many decisionmakers involved.

This report builds on a long National Research Council tradition of addressing the challenges that new demands for water pose in the West. In 1966 the Committee on Water, a forerunner of the Water Science and Technology Board, convened a special committee under the leadership of the distinguished student of water policy, Gilbert F. White, to explore the options available to improve water management in the Colorado River basin. That committee's report, *Water and Choice in the Colorado Basin: An Example of Alternatives in Water Management* (NAS, 1968), was a masterful survey of nontraditional alternatives for water management in the Colorado basin. The report called for more attention to the expanding equity and environmental demands on the river. It was an eloquent plea for policymakers to enlarge the range of options that they consider when formulating water resource policy. Specifically, *Water and Choice* advanced the then-heretical idea that states and the federal government should consider alternatives beyond project construction to stimulate regional growth and should give greater weight to environmental values.

Water and Choice accurately predicted many shifts in water resource policy that did occur in the past 25 years. The construction of new large-scale water projects is no longer central to regional development; the protection of instream uses and the promotion of social equity are now considered legitimate values. As a result, water managers, providers, and users must now seriously consider a wide range of management options such as water transfers to meet their objectives and to promote the twin goals of equity and efficiency. They are no longer theoretical options, as they once were.

Members of the committee would like to thank the many people across the West who met with us and helped us develop our thoughts. We talked with astute representatives of federal agencies, state governments, Indian tribes, environmental groups, agricultural groups, and many other interests. Although the committee takes sole responsibility for the ideas expressed in this report, it could not have completed its effort without this broad spectrum of help. I would also like to thank the project's financial sponsors—the Ford Foundation, the U.S. Department of the Interior, the U.S. Environmental Protection Agency, the East Bay Municipal Utility District, and the Metropolitan Water District of Southern California—for their courage in setting us free to debate such a challenging issue. Finally, I would like to express my profound thanks to the members of the committee. Each approached the task with enthusiasm and dedication. This report is a true peer collaboration and reflects consensus arrived at through mutual learning. The committee expresses special thanks to Wendy Melgin, who steered the project through the planning process before October 1989, and to Chris Elfring, who managed the project from the first case study to the final edit with dedication, humor, and a firm professional hand. Anita Hall and Wyetha Turney provided the staff assistance crucial to a project of this nature, and we thank them.

Some readers of this report may be disappointed that we do not offer a definitive resolution of issues such as which communities should be preserved from external change and which should be subjected to the full discipline of the market. They may be equally disappointed that we do not offer detailed federal and state legislative proposals to facilitate desirable transfers and discourage undesirable ones. Committee members have many different ideas on how to solve the problems addressed in this report, but our consensus was that at this stage of the transfer debate it would be presumptuous to put forward a menu of any great detail. Rather we hope that this report will shape the debate by highlighting the issues that state and federal governments must address and the possible avenues of approach.

A. Dan Tarlock, *Chair*
Committee on Western Water Management

Contents

APPENDIXES

Water Transfers in the West

Summary

Wise water management is among the most crucial challenges facing the American West. The normal flows of most western rivers are fully allocated, and new water supplies are scarce. Demand for water continues to increase as urbanization spreads. In addition, over the past 30 years the values placed on various uses of water have changed and the now urbanized West demands water not only for new municipal and industrial uses but also to ensure environmental stewardship. As never before, the public is becoming aware of the value of free-flowing streams and undammed canyons. As these resources—invaluable for fish and wildlife habitat, aesthetics, and outdoor recreation opportunities—become scarcer, the demand to preserve them intensifies. The relationship between water quantity and quality, long ignored, also is now a significant concern. These changing demands for water stress the primary water user in the West—irrigated agriculture.

The roots of irrigation go deep in the West, especially in the Southwest. When the mining, open range cattle, and dry farming economies were no longer able to sustain western settlement in the late nineteenth century, western promoters turned to irrigation. And irrigated agriculture remained the engine of western development until the 1970s. For almost a century, western water management was guided by five assumptions: (1) there should be easy private access to public water resources, (2) runoff should be captured and impounded for use during the dry growing season, (3) these captured waters should be used for multiple purposes, (4) water rates should be minimal for both agricultural and urban users, and (5) to settle the West and promote regional economic development, water resource development should be federally subsidized. In short, water management meant supply augmentation. But the era characterized by the construction of large subsidized water storage facilities and distribution systems—known as the reclamation era—appears to have ended, and an era of reallocation and improved management has begun. In his last public address, the late Scott M. Matheson, an astute student of western water policy and former governor of Utah, peered into the future and concluded that the "water policies of the nineteenth century no longer meet the needs of the twentieth century and will certainly not serve us well in the twenty-first century."*

Better management is imperative if we are to accommodate contemporary demands for a widening range of consumptive and nonconsumptive uses. This new thinking is already being implemented. Proposals to augment supply face stringent fiscal and political constraints. Instead of new projects, there is an increasing trend to transfer water from existing uses, primarily irrigation, to growing cities. Transfers also are beginning to be used to dedicate water to instream uses. With varying degrees of enthusiasm, many water users and influential segments of the environmental community have accepted the premise that water marketing—the transfer of water rights from existing to new uses at market value—should be a major component of future western water policy. This committee believes that voluntary water transfers are the most significant mechanisms available today for responding to the West's changing water needs but that broad "third party" participation is essential if transfers are to be both efficient and equitable. Third parties include those who hold vested water rights that may be at risk from a transfer, as well as those representing a range of economic, environmental, and social interests related to the transfer who claim a "nonproprietary" stake in the process.

The West has a long history of water transfers, which the committee defines as a change in the point of diversion or a change in the type or location of use, ranging from simple internal adjustments to actual sales. Any transfer can have significant third party effects depending on the scale, but the current interest in transfers is sparked primarily by the emergence of water markets. The idea that water is

*Matheson, S. M. 1991. Future Water Issues: Confrontation or Compromise? Journal of Soil and Water Conservation 46:96-97.

just another commodity to be allocated by the market has long been advocated by economically oriented policy analysts, but others have strongly disagreed. For a variety of reasons, recent years have brought an increase in the level of transfers in the West, and many of them are market driven.

The increased interest in water marketing results from a number of factors. Irrigated agriculture in the West uses large amounts of water. Lower prices for agricultural commodities, comparatively higher energy costs, and the problems of obtaining large amounts of capital all hinder the development of major new water supply augmentation projects for agriculture. There is no longer a strong constituency for subsidized federal water projects. Environmental concerns about irrigation-induced water quality problems are increasingly evident. In short, market forces and political forces have provoked a fundamental shift in values. Today, federal and state policymakers need to ensure that water allocation laws can respond to all water use demands—old and new—in an efficient and fair manner.

Water marketing has great potential to enhance the efficiency of water use. Markets respond to price signals to move resources from lower- to higher-valued uses. Markets respect existing property entitlements, and thus water rights holders set the pace of transition and receive compensation when water is transferred. Reliance on water marketing, as opposed to government subsidy and regulation, reflects a general societal belief that markets are a more effective way to allocate scarce resources to meet the twin goals of efficiency and equity. For sound reasons, water transfers will remain a central component of water policy in the next century.

However, there is a need for caution. Water markets cannot be expected to resemble more conventional markets for a variety of reasons, including the long tradition of subsidized water, the concentration of large blocks of water in public and private entities, and the equally long tradition that the use of water support the maintenance of a wide variety of public values. Transfers must thus be carefully evaluated because, as with any policy option, there are benefits and costs to their use. And significant costs—some concrete and others quite difficult to measure—can come at the expense of third parties.

The goal is not to promote transfers per se but to use them to accomplish better overall water management. To date, most policy analysis has concentrated on ways to lower the transaction costs of transfers. Often neglected, however, is the fact that transfers can impose significant third party effects that must be accounted for in any reallocation. If transfers are to achieve their potential, the decisionmaking process should bring all relevant third parties into the deliberations. This broad participation is necessary because water is a unique resource, different from other commodities, and markets alone cannot accurately reflect all the relevant values of this resource.

The focus of this report is on third party interests that are not sufficiently included in existing water allocation processes. This report characterizes the range of existing nonproprietary third party interests and describes the ways in which current water allocation institutions accommodate these interests. In addition to providing this general analysis of third party effects, the committee has examined a number of areas in the western United States where water transfers are occurring, or may soon occur, in an attempt to identify characteristics common to water transfers and to obtain firsthand knowledge about when they are considered "positive" and when they are potentially harmful to third parties. A comprehensive assessment of benefits and costs of water transfers is premature, because transfer theory exceeds transfer practice; thus the committee does not render definitive judgments about the role that water transfers should play in the future of western water allocation and how third party effects should be weighted by decisionmakers. Rather, the committee both acknowledges the merits of water transfers as a mechanism for meeting new demands and recognizes the legitimacy of a wide range of potentially affected third party interests in the transfer process.

The committee's basic conclusions are that allocation processes should accord third parties with water rights—and those without them—legally cognizable interests in transfers and that states should develop new ways to consider these interests. Water has never been allocated solely by markets, and market transfers are not an end in and of themselves but a means to the end of a water allocation process that serves both private and public interests.

An expanded set of criteria is needed to evaluate transfers and to accommodate the diverse and strongly held economic and cultural values associated with water use. Accordingly, in preparing this report the committee recognized the relevance of both economic techniques, which can be used to measure the value of water use and the costs of transfers, and other methods that permit more subtle and intangible values to be considered. The committee approached its study of water transfers with an optimistic sense of the role transfers can play in a new era of more efficient use. The committee concludes, however, that judicious intervention in water transfer processes will be necessary to avoid or ameliorate the adverse effects of some transfers.

In evaluating third party effects, the committee assumed that

• reallocation of water among uses will be a principal feature in a new era of western water management; increased conservation, increased use efficiency, and improved reservoir operation also will be essential;

• the general direction of reallocation will be from agricultural to municipal, industrial, recreational, and environmental uses;

• water markets involving willing buyer-seller transaction opportunities will continue to expand; and

• new formal and informal constraints on water transfers will be established until all parties are confident that the reallocation process includes consideration of all relevant interests.

THIRD PARTY IMPACTS AND OPPORTUNITIES

The term "third parties" is broad and includes everyone who is not a buyer or seller in a transfer negotiation. The general categories of parties who stand to be affected by transfers are (1) other water rights holders; (2) agriculture (including businesses and farmers in the area of origin); (3) the environment (including instream flows, wetlands and other ecosystems, water quality, and other interests affected by environmental changes); (4) urban interests; (5) ethnic communities and Indian tribes; (6) nonagricultural rural communities; and (7) federal taxpayers.

Third party impacts can stem from changes in the quality and quantity of water available for other uses, changes in the rate and timing of surface flows, and changes in ground water levels and recharge processes. Generally, these impacts are economic, social, or environmental in nature. Economic effects include impacts on incomes, jobs, and business opportunities. Social impacts include changes in community structure, cohesiveness, and control over water resources; such changes can occur in both rural and urban communities. Environmental impacts include effects on instream flow, wetlands and other ecosystems, water quality, recreational opportunities that are dependent on streamflows, and wildlife habitat.

Impacts can be both positive and negative, and assigning value to them is difficult. The underlying challenge of any process used to evaluate transfers is how to determine and balance equitably the relative benefits and costs. Techniques for measuring the impacts of water transfers are more precise for some types of impacts than others. It is difficult and unreliable, for instance, to apply economic measures to those impacts that are not usually measured in market terms, including most social, political, and environmental effects.

ASSESSING WATER TRANSFERS AND THEIR EFFECTS

To identify and study the impacts of water transfers on third parties, the committee developed a strategy to assess the characteristics of transfers and transfer opportunities. In particular, the type of transfer, primary motivation and process used, affected parties, and nature of the effects were examined (Table S.1). These factors were used to examine the nature of transfer activity in seven western areas:

- Truckee-Carson basins in Nevada
- Colorado Front Range-Arkansas River Valley
- Northern New Mexico
- Yakima basin in Washington
- Central Arizona
- Central Valley of California
- Imperial Valley of California

The case studies strive to (1) identify the incidence of third party effects, (2) identify those effects that were pervasive and those that were unique, (3) understand both the nature and the causes of third party effects, and (4) understand the actions available to mitigate or remedy any harmful effects. The committee's objective was not to judge the desirability of actions taken or not taken in a particular case. Rather, the committee used the analyses to highlight broad lessons about the nature, scope, and impacts of water transfers in general and to develop suggestions for improving the processes used to evaluate and regulate water transfer activity.

THE ROLE OF LAW IN THE TRANSFER PROCESS

Because water is seasonally and geographically limited in the West, encouraging the productive use of water has always been a key policy objective, from the days of the Anasazi to the present. Water use in the western United States, as in virtually all arid societies in the world, is regulated under rules designed to achieve broad public benefits. The prior appropriation doctrine—which gives the earliest user the right to take water from a stream and put it to "beneficial use" and to continue such use—was adopted during the settlement era as a way to allocate water so that it met both private needs and larger societal goals.

As western economies matured, the water rights system proved adaptable to increasing and competing demands. The key to adaptability was that water rights were not restricted to use on a particular parcel of land or to a specific type of use. In principle, rights could

TABLE S.1 Factors to Consider When Assessing Potential Water Transfers

Type of Transfer

- Change in ownership
- Change in point of diversion
- Change in use
- Change in systems operation
- Out-of-basin diversion

Primary Process for Transfer

- Voluntary
- Involuntary

Primary Market Forces for Transfer

- Government
  - Local
  - State
    - Executive
    - Legislative
    - Judicial
  - Federal
    - Executive
    - Legislative
    - Judicial

Affected Parties

- Rural communities
  - Support services
  - Erosion of tax base
  - Loss of natural resource base
- Agriculture
  - Remaining water users
  - Reallocation of rights
- Ethnic communities and Indian tribes
  - Ethnic communities
  - Indian communities
  - Agricultural maintenance and expansion
  - Other
- Environment
  - Instream flows
    - Recreation uses
    - Fish and wildlife
    - Hydroelectric power
  - Water quality
    - Damages to water users
    - Human health
    - Ecosystem effects
  - Ecosystem protection
    - Endangered species
    - Wetlands
    - Riparian habitat
  - Estuaries
- Urban interests
  - Intrastate transfer constraints
  - Tax-exempt status changes
- Federal taxpayers
  - National economic concerns
  - Windfall profits
- Other water rights holders
  - Junior rights
  - Senior rights
  - Loss of flexibility

Nature of Effects

- Economic (national/regional)
  - Lost revenue
  - Lost opportunities
  - New revenue
- Environmental
  - Instream/fish and wildlife
  - Recreation
  - Water quality
  - Wetlands
- Social
  - Rural communities
  - Municipalities
  - Other

be transferred from one user to another, and water could be delivered as far as technology and economies could move it. Such transfers were subject to the condition that a change in use must not injure any other water rights holder. States have enacted laws qualifying and complicating this simple perspective on the transferability of water rights but not altering the basic legality of this process.

The transfer of appropriative rights has always been possible, but it is only recently—in the era of full appropriation of many western streams—that reallocation has become the main source of water for new demands. States and their citizens are realizing, however, that voluntary transfers among private parties may affect an array of interests that are not adequately protected by the laws and processes that govern transfers. Western states, tribal governments, the federal government, and water districts all have opportunities and responsibilities to deal with the effects of transfers. The most direct way to do so is through policy planning, impact assessment, and public interest review of proposed projects. The more comprehensive and predictable the review process, the more incentive water sellers and buyers will have to accommodate these interests throughout the transfer process.

In addition, several existing laws offer ways to address the effects of transfers. State instream flow laws can be important for limiting the environmental and economic harm sometimes caused by transfers. State water quality protection goals can be furthered in the transfer process if statutes and procedures are clarified to specify that purpose. Federal project water can be used in ways compatible with state instream flow laws, with transfers allowed only when these state laws are satisfied. Ultimately, protection of instream flows will depend on acquisition of senior rights.

CONCLUSIONS AND RECOMMENDATIONS

The recognition and protection of third party interests are essential if water transfers are to achieve their potential to reallocate water to meet new demands. Transfers represent both a break with past western water allocation practices and, at the same time, a continuation of past practices. Transfers can bring the benefits of the market to a system that has often subordinated efficiency to distributional concerns. But the West has never treated water as just another commodity and should not do so now. There must be a balance between efficiency and fairness. Each jurisdiction must devise its own laws and processes to achieve this balance.

The committee believes that many of the problems and opportunities illustrated in the various case studies will recur throughout the West. Thus valuable lessons for future federal and state water allocation policy can be drawn from the studies. To this end, the committee offers the following conclusions and recommendations (which are discussed in more detail in Chapter 12).

Water Transfer Opportunities

Water transfers can promote the efficient reallocation of water while protecting other, water-dependent values recognized by society. All levels of government should recognize the potential usefulness of water transfers as a means of responding to changing demands for use of water resources and should facilitate voluntary water transfers as a component of policies for overall water allocation and management, subject to processes designed to protect well-defined third party interests. Third party protection should be seen not simply as a constraint but as a legitimate component of the transfer process.

State and Tribal Authority

State and tribal governments have primary authority and responsibility for enabling and regulating water transfers, including identification and appropriate mitigation of third party effects. State and tribal administrators should develop and publish clear criteria and guidelines for evaluating water transfer proposals and addressing potential third party effects. State and tribal administrative processes should provide for public and broad third party representation in the review of water transfer proposals. In addition to normal actions such as notices of proceedings, public hearings, and protest opportunities, programs should also include affirmative review of potential third party impacts in cases likely to involve significant effects. State and tribal processes should seek to regulate water rights transfers in ways appropriate to the scale of effects with the dual objectives of avoiding excessive transaction costs while providing meaningful consideration of third party interests.

State laws should allow governmental entities to acquire water rights for instream flow purposes with the same priority and protection against injury enjoyed by rights held for other uses. Affirmative state policies may be necessary and appropriate to acquire sufficient instream flow rights to mitigate the effects of historic diminutions of streamflow. States should provide leadership in exercising their water administration and planning responsibilities to identify opportunities for water transfers that might serve as instruments for achieving a wide range of water management objectives.

The Costs of Transfers

Water transfer law and policies should be designed to consider the interests of the trading partners, third parties, and the environment in a cost-effective manner. The costs of mitigating third party effects should be internalized as a cost of the transfer—that is, the beneficiaries or proponents of the transfer should bear the mitigation costs as a matter of law and equity. Therefore the cost of the transfer should include sufficient funds to help mitigate third party effects, in the form of water, money, or other compensation.

To help reduce transaction costs, policies might be designed so that, in general, transfers of acquired rights are limited to consumptive use. This may entail setting state, river basin, or regional standards for the consumptive use of water per unit of irrigated land, based on crop type, historic water availability, and other local variables. Such standards should be flexible enough to account for variations in water availability and local conditions. Third parties should not have to develop data on the transferable quantity; data should be developed by the buyer or seller. Regulatory requirements should be designed to encourage negotiated resolutions of conflicts. Consideration should be given to processes (e.g., a state water court) other than judicial proceedings to provide the initial evaluation of transfer proposals.

Area-of-Origin Impacts

Water transfers between basins should be evaluated to determine and account for the special impacts on interests in the areas of origin. States and tribal governments should develop specific policies to guide water transfer approval processes regarding the community and the environmental consequences of transferring water from one basin to another because such transfers may have serious long-term consequences.

Water transfer processes should formally recognize interests within basins of origin that are of statewide and regional importance, and these interests should be weighed when transbasin exports are being considered. Although each state or tribe should select the approach that suits its needs best, area-of-origin protection generally would include impact assessment, opportunities for all affected interests to be heard, regulatory mechanisms to help avoid adverse effects, compensation (e.g., financial payments or mitigation), and authority to deny a proposed transfer or water use involving a transbasin export if the effects are judged to be unacceptable. States should revise laws that now exempt water facilities from taxation by the county of origin either because the exporter is a public entity or because of provisions that make such facilities taxable only in the county where the water is used. Mechanisms to compensate communities for transfer-related losses of tax base, such as an annual payment in lieu of taxes, may be needed.

The Public Interest

Public interest considerations should be included among the third party issues and legal provisions for permitting, conditioning, and denying water transfers. To protect third parties in water rights transfers, the public interest language in western states' water laws should be reviewed, clarified, and, where appropriate, more vigorously applied. States should develop definitions and criteria for assessing what constitutes the public interest, perhaps benefiting from the legislative and judicial experiences of the states of Idaho and Alaska. Such definitions should embrace existing water rights holders, environmental water needs for ecosystem protection, and social and cultural values in basins of origin.

To the extent that public trust concepts and values cannot be dependably represented under existing laws and policies, states should develop new laws, institutions, and administrative tools for doing so. Key elements of public trust administration might include comprehensive planning at the river basin level, including the identification of existing social and environmental values dependent on water; clearly defined procedures to guide applicants seeking water transfer approval; and institutional arrangements for holding and managing water for instream and environmental uses.

Environmental Impacts

Environmental impacts can and should be considered by state, tribal, and federal agencies when potential water transfers are evaluated. Federal, state, and tribal water transfer policies and laws should ensure consideration of ecological values affected by transfers, for example, the goals of protecting riparian and wetland habitats and the water needs of endangered species. Adjustment of water laws, administrative practices, and water supply strategies may be necessary to achieve this end. States should develop inventories of wetland and riparian systems that depend on surface water systems likely to be sought for transfers so that transfers can then be evaluated in light of ecosystem protection and restoration goals.

When water is transferred to new uses in new areas, there is an opportunity to promote overall efficiency by dedicating some of the newly available supply to public uses such as environmental protection. Congress should consider using a share of the receipts gained from marketing federal project water to acquire water to protect western biota and habitats. The federal Land and Water Conservation Fund Act might be amended to permit both state and local grant recipients and federal agency participants to use fund money to acquire water rights for environmental protection and mitigation.

The Unique Interests of Indian and Hispanic Communities

Traditional Indian and Hispanic communities have unique interests relating to water transfer policies because long-established water uses are often central to the survival of their cultural identity. These interests merit special consideration when proposed transfers are evaluated. States should carefully scrutinize any proposed water transfer that could adversely affect water supplies for Indian and Hispanic communities. States should consider enacting legislation that permits the establishment of historical or cultural zones as a means of insulating these communities from further injury. Transfers that would have adverse impacts on these zones would receive strict scrutiny.

Transfers on Indian Reservations

Tribal governments should consider special factors in approving and administering water transfers on their reservations. When Indian water rights are transferred or applied to off-reservation uses, tribal governments should establish procedures to evaluate third party effects. State and federal governments should cooperate with tribes in evaluating and implementing mutually beneficial transfers.

Water Salvage Laws

Water laws should be enacted to promote water conservation and salvage while protecting third party interests. States, tribal governments, and federal agencies should establish programs to reduce the uncertainties and costs involved in water conservation and salvage, by facilitating and providing incentives to encourage conservation and reuse of water. In most cases, conservation programs will not have the adverse impacts that transfers may have. Nevertheless, state, tribal, and federal water transfer review processes should take full account of the third party effects of a transfer of conserved or salvaged water, just as they should with any other transfer.

Water Quality-Water Quantity and Surface-Ground Water Interrelationships

Water transfer reviews should consider the interrelationships between water quality and water quantity and also between surface and ground water resources. States and tribes should accelerate the development of laws, policies, and administrative procedures to ensure consideration of water quality changes that might result from transfers involving the use of municipal effluent and other water of impaired quality. States and tribes should encourage conjunctive management of surface and ground water, based on knowledge of hydrologic connections and a full consideration of the related water quality issues and other potential third party effects. States and tribes should require agencies to develop technical capabilities for evaluating and monitoring surface and ground water quality as part of the transfer evaluation process.

Federal Policy

Federal reclamation law makes inadequate provision for the transfer of federal project water because it was assumed that the water would always be used within the project. Given changing demands, the federal government should now have a specific transfer policy. Federal legislative and administrative policies should more clearly support federal water transfers while paying careful attention to third party effects and to the distribution of benefits from transfers involving federal project water. The Secretary of the Interior should develop a formal process for assessing transfers of federal project water. This process should identify the role of those agencies of the U.S. Department of the Interior responsible for natural resource stewardship and should allow for consultation with tribal governments, states, and the interested public to ensure that major third party effects are assessed and mitigated.

Congress should set clear policy for the distribution of profits from the resale of federal project water, and this policy should provide a portion of the economic value in federal project water to be recaptured for public use. The Secretary of the Interior should require greater consistency in applying the department's general policy supporting voluntary transfers and should specify that there is authority to transfer water among as well as within project service areas.

BALANCING EFFICIENCY AND EQUITY

The allocations of water produced by the doctrine of prior appropriation represent the prevailing social values of an early era that favored development of mining and irrigated agriculture. However, as social values change, these allocations are sometimes perceived as misallocations. Throughout its investigation of the nature, scope, and impacts of water transfers, the committee found much evidence of what might be called a "clash among values." These conflicts are felt more sharply today than ever before because of increasing competition for a limited resource. In many of the cases studied, the clash is between the older, rural, agricultural West and the newer, more urbanized West. In other instances, the clash is between off-stream and instream uses of water—the result of a more environmentally conscious West.

As transfers become common in western basins, the necessity of comprehensive planning and management will become clear. Water management institutions should develop the technical capability to assess the full range of impacts associated with potential water transfers and should weigh the merits and problems broadly along with other available management options and the effects of each.

There are inherent limitations in the capability of market mechanisms to deal with nonmarket goods and externalities, and these must be addressed through institutional change. The committee concludes that state and tribal governments must work to give third parties a more effective voice in the water transfer decisionmaking process.

The cases examined by the committee raise a diverse set of questions about effective consideration of third party impacts. In northern New Mexico, for instance, the special needs of a historic ethnic community with water allocation institutions that predate Anglo settlement are not addressed by the existing state system. The Arizona case reflects the conflicting needs and aspirations of urban and rural populations. The Truckee-Carson, Nevada, example illustrates an exceptionally diverse array of interests—two Indian tribes with differing cultural and environmental needs, a growing urban area with increasing water requirements, and more than one natural wetland area needing water to maintain migratory waterfowl and other fish and wildlife values.

Designing mechanisms to accommodate third party interests is difficult, almost inevitably imposing new burdens on both the transferring parties and the administrative agencies that review water transfers. The quest is for a balance point that allows broader participation in decisionmaking while not inhibiting desirable transfers. The recognition that third party interests are legitimate is a first step toward this accommodation.

1

Pressures for Change

The West is defined . . . by inadequate rainfall. . . . We can't create water, or increase the supply. We can only hold back and redistribute what there is.

Wallace Stegner, 1987

The American West faces many challenges, but none is more essential than the wise management of its limited water resources. Many of its river basins are fully allocated, yet the demand for water is increasing in response to continued population growth and the emergence of new social values.

In the West today, the era characterized by the construction of large subsidized water storage facilities and distribution systems has ended, and an era of reallocation and improved management has begun. This policy shift reflects the region's adjustment to changes in society's demands and values. An urbanized West demands water for new municipal and industrial uses as well as water to ensure environmental quality and recreational opportunities. Western water policies are based increasingly on the need to manage water resources better to accommodate a wide range of interests. The trend is to move water from existing uses, primarily irrigation, to growing cities, which can no longer secure additional supplies through surface supply augmentation or ground water pumping, and to a variety of instream uses.

There is an increasing consensus in the western water community that new demands must be met primarily through the reallocation of existing supplies. The use of voluntary water transfers, or water marketing, has been accepted by many members of the western water community—sometimes with great enthusiasm and sometimes with resignation—as the best method for reallocating existing supplies to these new uses (Brickson, 1991; Wahl, 1989; Willey, 1985).

This report examines the issues that arise out of the western states' increased interest in water transfer mechanisms for reallocating water in an efficient and fair manner. Water transfers increasingly are seen as an important management option because they present opportunities to meet municipal and industrial demands, bolster environmental and recreational values, and shift water to new uses with minimal disruption to existing rights holders. Reliance on markets as opposed to government subsidy and regulation reflects a general societal belief that markets are a more effective way to allocate scarce resources efficiently and fairly. But many questions about the impacts of water transfers remain unanswered, especially regarding their effects on a wide range of third parties.

To date, many discussions have centered on describing transfers and the institutions that govern them. For the past decade the policy focus has been on how states and the federal government can create the necessary incentives to reduce associated transaction costs and to encourage more water transfers. The focus in this report is different. The committee acknowledges that water transfers can produce many benefits, and it documents instances where these benefits appear to have been generated. But the special emphasis here is on the potentially neglected third party effects of transfers—the impacts these changes in water allocation and use can have on the people, communities, and environments that are not typically considered parties in the transfer process. These effects are examined primarily in relation to the transfer of water from one use to another, but the committee also discusses other forms of reallocation that can result from judicial declarations of rights and new management initiatives when these actions relate to water transfers.

Third parties are both those who hold vested water rights that may be at risk from a transfer and those who claim a "nonproprietary" stake in the process and represent a range of economic, environmental, and social issues related to the transfer. These third parties—perhaps more correctly called "affected parties"—include areas of origin; Indian tribes; other minority cultures (primarily Hispanic) that want to preserve traditional cooperative land and water use patterns; communities that depend on irrigated agriculture or water-based recreation; boaters; anglers; and broad segments of the public who care about wetlands, riparian areas, endangered species, instream flows, and other environmental values that might be harmed or enhanced by a change in water use.

This report is especially concerned with third parties whose interests are not now adequately represented in the institutions and processes of water allocation. It characterizes the range of existing nonproprietary third party interests and describes ways in which current water allocation institutions accommodate or neglect these interests. In addition to this general analysis of third party effects, the committee has examined seven specific areas in the West where water transfers are occurring or may occur in the foreseeable future. These cases illustrate that water transfers—despite their often positive benefits—also have the potential to impose significant harm on parties outside the transfer process.

It is premature to offer a comprehensive assessment of the benefits and costs of water transfers, because transfer theory exceeds transfer practice. Thus the committee does not render definitive judgments about the role that water transfers should play in the future of western water allocation or how specific third party effects should be weighted by decisionmakers. Rather, the committee endorses the merits of water transfers as a mechanism for meeting new demands and recognizes the legitimacy of a wide range of potentially affected third party interests. This committee believes that allocation processes should accord third parties (both those with water rights and those without them) legally cognizable interests in transfer discussions and that state and tribal governments should develop new ways to consider these interests. Western water has never been allocated solely by markets, and market transfers are not an end in and of themselves. Rather, they should be part of an allocation process that serves both private and public interests.

Western water law is currently moving in two directions. To encourage transfers, states are trying to reduce the transaction costs incurred when a transfer is proposed and to create greater incentives to conserve and perhaps transfer conserved water. At the same time, much more attention is being paid to new water uses, often in situ, and to the broader consideration of impacts on third party effects. These two trends can have opposing effects on transaction costs.

Despite a potential increase in cost, the inclusion of these new voices in water transfers is a logical extension of the western states' long insistence that water use promote broad societal goals. The range of interests represented at the bargaining table must be broadened if water transfers are to achieve their potential. Otherwise, these third party effects act only as constraints on reallocation and thus further reduce the incentive to use transfers in appropriate situations. The challenge for water regulators and providers is to devise processes that encourage transfers with real benefits and restrain or

Why the Committee Focused on Transfers Within States Rather Than Between States

Water can be transferred within a state or across state and major river basin boundaries. This report focuses on the simpler of the two scenarios: intrastate water transfers. This type of transfer occurs far more frequently than transfers between states or between interstate basins, and intrastate transfers often raise similar issues for water resource officials throughout the West.

Interstate transfers, on the other hand, typically introduce unique political and legal constraints and considerations. Individual states have adopted a number of different legislative strategies in attempts to safeguard their interests in the context of proposals to transfer water out of state. For example, northwestern state delegations have long worried about southwestern designs on the Columbia River and have successfully pushed through Congress a prohibition on federal funding of studies to examine proposals to transfer water from the Columbia basin without the consent of basin governors. Interstate transfers on the same river can be equally controversial because they pit states with uneven growth rates against one another. Proposals to transfer water from the upper Colorado River basin to the lower basin result in substantial controversy about the "law of the river," the phrase used to describe the numerous federal laws, court decisions, and interstate compacts dictating allocation of the Colorado River.

Because of these unique legal and institutional constraints and the large scale of major interstate, interbasin transfers, such transfers are relatively rare; thus an attempt such as this to draw lessons from our past experiences with water transfers is best focused on intrastate efforts. The applicability of these lessons to interstate transfers may be limited by the factors discussed above.

condition those that impose high costs on legitimate third party interests. Each state and Indian tribe will have to decide how to strike this balance. In this report, the committee offers some general principles and suggestions that might help accomplish this objective.

Water marketing strategies build on the fact that western water rights have long been classified as property rights that exist independently of land ownership. Much of the West's urban and agricultural development has depended on transfers from one watershed to another, and water rights have long been transferable. However, water rights transfers were historically a secondary means of ensuring adequate supplies, compared to supply augmentation using ambitious construction projects. The legacy this tradition left to western water law is that water rights are transferable property rights, but such transfers can be difficult to accomplish because of the unique nature of water rights. Water rights are more contingent compared to property rights in land. They are dependent on the dedication of water to productive use and they are inherently correlative. A water right is valid only so long as the water is put to a beneficial use and all rights are defined in relation to other users.

In short, a water rights holder only acquires the right to *use* a specific quantity of water under specified conditions. The limited nature of property rights in water is captured by the legal concept that all water rights are usufructuary. A water transfer cannot take place if other water rights holders—even junior rights holders—will be injured. Thus every water transfer is potentially a multiparty transaction. The traditionally protected third parties—water rights holders—no longer represent the entire class of relevant interests with a legitimate stake in transfers.

Most western states now have some procedures to protect third parties, but existing water transfer institutions accommodate these nontraditional and new interests with varying degrees of success. Some procedures fit into the basic appropriation scheme. For instance, junior appropriators have standing to object to a transfer, and Indian tribes can assert their generally superior reserved rights. In addition, in Colorado, Idaho, Nevada, and Wyoming, a state agency may hold instream flow appropriations, and most states require that the state engineer consider the public interest in both new appropriations and transfers. However, the committee's case studies illustrate that the list of third party interests that are legally protected is shorter than the list of interests actually affected. Consequently, the task of determining how to assess and accommodate such concerns has just begun.

The challenge today is to satisfy existing demands while dealing equitably with rural communities, the environment, Indian tribes, and ethnic communities. In this report, the committee suggests a number of elements that should be considered in evaluating transfers and recommends changes in existing state and federal laws and policies to promote transfers that reallocate water without adverse effects. Tribal governments may also find these ideas useful as they develop water laws and policies for the Indian reservations. The committee also identifies situations where transfers need to be reviewed and either restrained or reworked to mitigate adverse effects.

There is no simple answer regarding the size of the negotiating table for water transfers. As Joseph L. Sax, a leading water law scholar, observed to the committee, the West must accept the reality that water transfers will more often resemble diplomatic negotiations than simple market transactions. The committee does, however, endorse the position that in the long run expanded efforts to consider third party interests will promote both more efficient and more equitable allocation of western water. Added short-run costs are a necessary price to pay for the transition from a water policy based on supply augmentation to one based on reallocation.

THE HISTORICAL CONTEXT

Water transfers are a signal of a change in the culture of western water allocation, and this context defines how the benefits and costs of transfers are judged. Water management is moving from an era of development to an era of reallocation and more intensive management of the resource as competing demands intensify.

Since the reclamation era of the late nineteenth and early twentieth centuries—a time characterized by the construction of large water storage and distribution systems—water resource policy in the western United States has been guided by five assumptions: (1) there should be easy private access to public resources, (2) spring runoffs should be captured and impounded for use during the dry growing season, (3) these captured waters should be used for multiple purposes, (4) water rates should be minimal for both agricultural and urban users, and (5) to settle the West and promote regional economic development, water resource development should be federally subsidized. The reclamation era was characterized by the acquisition of private rights in public waters. It was a time when water conservation meant the construction of storage and distribution facilities. This historic policy is now being replaced with a more diverse one (Leopold, 1990) that recognizes the need to accommodate a wide range of consumptive and nonconsumptive water uses, primarily through better employment and reallocation of existing water supplies and pricing standards that reflect the full cost of the use of water.

The roots of irrigation go deep in the West, especially in the Southwest. Initial efforts were cooperative: the Hohokam Indians developed a complex communal irrigation society along the lower Gila River in Arizona (Dozier, 1970), and the Anasazi continued the tradition until the fourteenth century, when they abandoned their villages. Spanish colonists in New Mexico and Texas blended their own irrigation culture, a legacy of the Moors (Gibb, 1960), with indigenous practices. In the nineteenth century, the Protestant missionaries who settled the Pacific Northwest used ditch irrigation. The major irrigation colonizers, however, were the Mormons, who began irrigating immediately after they entered the valley of the Great Salt Lake. Other early pioneers in Colorado and California formed irrigation colonies, which were originally part of the collective utopian movements of nineteenth-century America (Dunbar, 1983).

These early irrigation efforts, however, were small. Large-scale western settlement in the nineteenth century required a larger scale of activity, and formal institutions were needed to apply strict priorities among rights to water use to compensate for the seasonal limitations of the climate. These institutions historically adapted to chronic water scarcity and unreliability by seeking outside investment capital to support construction of water management facilities—capital that could be obtained only if the legal rights to the water to be developed were secure. After the mining, open range cattle, and dry farming economies were unable to sustain western settlement and development in the late nineteenth century (Webb, 1931), western promoters turned to irrigation to settle the West. Soon, irrigated agriculture—and the values it promoted—were supported by state laws and federal monies throughout the West, and support for an irrigation-based water ethic continued for the next century.

The law responded creatively to the development of a water-dependent society in the West by allowing the easy creation of private rights in public resources. This trend began during the first wave of settlement from the East—the Gold Rush—when the demand for water to work placer claims exceeded dry season streamflows in the booming mining camps (Leshy, 1990). To allow the early miners to reap the fruits of their enterprises, the unique legal doctrine of prior appropriation—providing that the party who first places water to beneficial use has the right to withdraw as much water as he can put to that use, ahead of any other user from the same source—was adopted by the courts and then by state legislatures and implicitly ratified by Congress.

Prior appropriation permitted the assignment of exclusive property rights in common resources. The security of these rights, based on priority of use, was later needed by developers of large irrigation systems raising capital to build storage and distribution facilities. Prior appropriation alone could not provide the necessary security for the scale of federal, state, and private investment needed to keep the West wet enough to prosper. Large water storage systems were necessary to overcome annual and seasonal variability in streamflows, and conveyance systems to deliver water to distant points of use followed logically from prior appropriation. The 1868 Powell Survey and subsequent federal water resource surveys contributed the idea that rivers should be impounded for irrigation, flood control, and hydroelectric power generation. "The greater storage of water must come from the construction of great reservoirs in the highlands where lateral valleys may be dammed and the mainstreams conducted into them by canals" (Powell, 1962).

Initial western irrigation projects were sometimes poorly financed, and many of them failed. In the early 1890s, there were only 3.5 million acres under irrigation, mostly concentrated in Utah, southern California, and eastern Colorado. Irrigation then became a national crusade, and the West began a period of rapid, sustained growth and prosperity. Between 1890 and 1910 the populations of Idaho and Nevada doubled, mainly because of irrigation. "By 1910, reclamation officials were pointing to the miracles already wrought in the desert—to cities that had sprung from the sagebrush and to a new land of fortune and opportunity that had materialized in the desert at the wave of the magic wand of water" (Athearn, 1986).

Together, the prior appropriation doctrine and the water development supported by federal spending produced the irrigation and urban oasis economy that characterizes much of the West today. Rights to use these augmented and more dependable supplies of water were held by public entities (either the United States or special districts) under the prior appropriation doctrine and shared by contract or subcontract with individual users. In this way a system of water allocation originally conceived in terms of private property rights became controlled by government entities.

The West's water ethic is now changing, and with it the nature of water demands. The list of principal water uses has expanded beyond irrigation, municipal and industrial supply, and hydroelectric power generation to include recreation and environmental quality. The successful federal effort to settle the West through reclamation is evolving to reflect changing national priorities. We are moving from an era premised on the continual development of new supplies to a reallocation era premised primarily on the better use of existing supplies.

WHY WATER TRANSFERS OCCUR IN THE WEST

That water is a limited resource in the arid West becomes more apparent as the range of demands increases. Water initially was allocated simply by putting it to a beneficial use, mostly irrigation; the doctrine of prior appropriation was indifferent to the economic value of the use. However, as competition for water has increased, economists and others have argued that water should be transferred "from low-value irrigation use to high-value noncrop use" (National Academy of Sciences, 1968). The classic rationale for all economic activity—gains from trade—motivates most water transfers. Buyers perceive that the cost of purchasing existing water rights and transferring water to new locations, seasons, or purposes of use is less than the cost of alternative means of securing needed supplies. Conversely, sellers—generally farmers—sell when the price offered is greater than the economic value of the crops or livestock they produce. The net result is that the new use generates higher economic returns than the old use.

The economic theory of water transfers is simple, but for transfers to occur, two conditions must be met. First, the benefits to buyers must be great enough (or be perceived as great enough) to outweigh the costs of obtaining water by alternative sources or by reduced demand plus all the transaction costs. Second, the costs of buying or leasing water, which can include political costs and legal uncertainties, must be less than the costs of other means of obtaining water—such as contracting for water deliveries from a public water project or a technological approach such as desalinization.

Recent studies of western water transfers suggest that more transfers of water are occurring now, and in more areas, than ever before. Data on applications filed for water rights transfers in 17 western states between 1963 and 1982 indicate a substantial increase in transfer applications (MacDonnell, 1990). There are several reasons why the level of interest in water transfers is increasing. With the exceptions of Montana, Wyoming, and the Great Plains states, most western states have experienced rapid rates of population and economic growth since World War II, especially in the last decade. More importantly, regional economies have shifted away from agriculture, livestock, and mining toward a mix of industrial and service jobs similar to that elsewhere in the national economy. Irrigated agriculture remains the predominant water user in the West, but the nonagricultural sectors now provide the majority of jobs and income in most states. The construction, manufacturing, tourism, service, and government sectors of the western states now rival agriculture in economic importance.

In addition, the shift in economies and lifestyles that has come with the latest wave of western settlers has fueled increased environmental demands for water to maintain streamflows to protect fisheries, support wildlife, and maintain riparian corridors—all scarce resources in the arid West. Water quality concerns also have gained attention. Finally, the drought years of the late 1980s have prompted many cities to become more aggressive in seeking additional supplies to protect their populations from future shortages. The uncertainties offered by global climate change scenarios reinforce this drive to increase the margin of safety in areas vulnerable to drought. All of these trends encourage water transfers.

Continued reliance on supply augmentation would be a difficult way to satisfy these new demands. The federal government has begun to frown on irrigation subsidies. Most streams are fully appropriated, and it is increasingly common for some portion of the flow of streams to be dedicated to instream values. Arizona, Idaho, New Mexico, and some other states have set severe limits on new ground water pumping. The costs of developing new water supplies have risen for several reasons: the best reservoir sites have been used, environmental considerations and conflicting water claims have prompted litigation resulting in project delays and costly impact studies, and the federal government is less willing to subsidize project costs. In short, the appropriation of new water rights and the construction of new storage and distribution facilities are no longer the most likely or the least costly alternatives open to water providers in most cases. These changes make it more likely that the second condition for a water transaction will be satisfied—that a transfer will be a less costly way to satisfy new demands than the alternatives.

CHANGING DEMANDS

Competition for water has always existed in the West, but this competition was easier to address in the past, when the major competing water interests all held water rights. Agriculture, hydroelectric power, industry, and growing cities have long required water, but all these purposes could be accommodated by the prior appropriation system, and they were able to share available supplies with remarkable harmony. For instance, in many cases hydroelectric generators used the water first and passed it downstream to farmers and cities. Water rights were pooled to support large-scale irrigation. As it became necessary to irrigate larger tracts and to supplement privately financed systems, farmers were willing to subordinate their rights to be used under federally financed projects. Government-subsidized projects also provided most of the West's hydropower.

Municipal and industrial growth was based initially on appropriating modest new rights to water in areas where water had not been developed intensely. The increasingly heavy urbanization of the West, however, made it necessary to transport water from distant watersheds where new appropriations were possible. In those streams where most of the water was subject to existing rights, new uses depended on the transfer of existing senior irrigation rights. Agricultural water rights moved to municipal uses in spite of legal impediments and often high transaction costs. Transfers were not, however, the major means of meeting new demands. Municipal water providers were motivated by restrictions and the political unpopularity of agricultural transfers to pursue the even more costly alternative of transbasin diversions.

Irrigated agriculture is stable at best in some places in the West and is under stress or in decline in others (Solley et al., 1988). Cities and industries continue to grow, and they are willing to pay high prices for water. They seek to build major storage facilities and acquire rights from farmers whose uses command lower economic value. These changes require sales and leases of existing rights and often entail the transport of water from one region, typically out of a watershed, to another. As a result, communities dependent on existing uses sometimes decline in relation to those that gain the new use of water.

The most remarkable new demand is for water in place. Recreation has become a leading industry in the West, now rivaling or surpassing agriculture or mining in economic importance in nearly all western states. Boating, fishing, hunting, and other outdoor enjoyment all require abundant good-quality water in place. Although techniques to assess the economic contributions of in-place water are imprecise, in many instances the value is considerable. Apart from the substantial economic value derived from water in place, there is a growing public demand for maintenance of ecological integrity. This new view is based on a better understanding of natural systems, and it is bolstered by esthetic and emotional values. Changing water uses from consumptive off-stream uses to instream uses is possible under the laws of some states but difficult economically and politically. It is politically difficult because the diverse proponents of instream flow tend not to be organized into coalitions with the political power or funds to negotiate the transfers of senior rights.

Public agencies and officials have focused most studies to date on the efficacy of western water law as a device for facilitating transfers of water. Recommendations are typically aimed at removing obstacles to the marketability of water rights. But as the attractiveness of water transfers becomes more widely understood, inadequacies in the laws and policies regulating transfers also become more obvious. Existing law in some states still tends to impede desirable transfers, although these impediments are being removed. The greatest problem, however, is that there are interests that deserve protection from the negative effects of transfers but that water law was designed primarily to protect water rights holders, not other interests that might be affected.

SOME RECENT TRANSFERS

The motivations for transfers can vary. Interest in transfers was stimulated during the energy boom of the 1970s. Studies predicted that water scarcity would inhibit mineral extraction and power production, and energy companies and utilities began purchasing water rights from all available sources. The predictions were largely wrong; nonetheless, many farm-to-industry sales occurred around the West. During occasional downturns in the agricultural economy, some farmers found that they could make more money by selling their water rights than by raising irrigated crops. Buyers have been primarily urban developers and municipalities who need additional supplies for growth. Environmental interests are also purchasing water to transfer to instream flows to restore fisheries and wetlands and to develop or maintain recreation-oriented local economies.

Some investors have speculated that the price of water rights would rise, and several groups have bought water rights in recent years. Sometimes buyers lease the water back to the irrigators who sold it, which lets farming continue until the water is actually needed to accommodate new urban growth. So far, however, there has been little market demand for these pooled rights held by investors. Mining interests, energy companies, and other industries are not currently major water buyers, owing to slow growth in these sectors, but they could enter the market again if energy and mineral prices rise or the federal government takes new initiatives to reduce our dependency on imported oil. New vitality in these sectors would stimulate additional interest in water transfers.

Tens of millions of dollars were spent on western water rights during the 1980s, primarily in the Southwest. (The prices reported in the next few paragraphs refer to the purchase of water rights—perpetual access to a specific quantity of water—except where the text specifically notes that the transaction is a lease of water or an arrangement for a one-time use of water.) In Colorado, cities and developers along the Front Range (Denver, Colorado Springs, and Fort Collins) have purchased agricultural water rights to meet growing water demands. Prices in the late 1980s ranged from $1,000 per acre-foot ($810 per megaliter (ML)) of water rights to more than $5,000 per acre-foot ($4,050 per ML) in some areas that have few alternative

Water as an Investment

Investors have been active in western water markets for a number of years. Investors purchase water rights—like other commodities—in the expectation that they eventually will be able to sell them to cities, industries, and private developers at a profit. This activity increased in the late 1980s, especially when a dramatic stock market decline in 1987 pushed investors to analyze alternative investments while widespread drought focused national attention on the long-term value of reliable water supplies. A number of popular publications, including *Newsweek, Business Week,* the *Wall Street Journal,* and the *New York Times,* ran articles on water marketing and investing in water rights. Groups of investors pooled money to purchase a diverse portfolio of water rights.

Such portfolios strive to acquire ground water entitlements, surface rights, irrigation district shares, reservoir stock, and other water entitlements that appear to be secure water rights and that are expected to appreciate in value. Several investment funds also have been created to acquire water rights. Prudential Bache formed a limited partnership to raise $20 million for water rights investments, offering shares at $50,000 a piece to qualified investors with a 10-year investment period.

Another investment fund, Western Water Rights Management, Inc. (WWRM, Inc.), raised $35 million to invest in water rights, based on funds provided largely by East Coast and foreign investors. The water acquisitions can be resold to developers and municipalities who seek to lock in reliable water supplies to accommodate development projects and population growth over the next decade. WWRM, Inc., acquired numerous water rights in Colorado's rapidly growing Front Range and is analyzing acquisitions in several other areas.

The Resource Conservation Group, Inc., was formed to promote transfers of underused water in Colorado, Utah, and Wyoming to rapidly growing areas in California, Arizona, and Nevada. The investors have retained four former governors of southwestern states to facilitate proposed interstate water sales. Western Water Development, Inc., based in the Reno area, and American Water Development, Inc., based in the Denver area, are two other multimillion dollar funds specifically designed for water rights investment.

Although investment groups such as these have supported the price of water rights in the West through their acquisitions of inventories, they have engaged in very few profitable sales as yet.

supplies.[1] In Arizona, water rights sold for more than $1,500 per acre-foot ($1,210 per ML) in the Phoenix and Tucson areas. "Water ranches"—agricultural lands purchased solely for the associated water entitlement—total tens of thousands of acres. Investors and cities who have bought water ranches eventually plan to move the water from its present uses to provide for urban growth (Colby, 1990b).

The Truckee and Carson river basins, near Reno in western Nevada, also have an active market in water rights. Senior irrigation rights sold in the 1980s for between $2,000 and $3,000 per acre-foot (between $1,610 and $2,430 per ML) to developers and municipal water providers and, most recently, to environmental groups wanting to restore fragile wetlands. Water rights prices are lower in central Utah, where urban growth and the need for new supplies are not so great. More than 100,000 acre-feet (123,400 ML) of water rights were purchased in the late 1980s in the Salt Lake City area at prices between $160 and $250 per acre-foot ($130 and $200 per ML). For water rights around Albuquerque, New Mexico, where urban growth has stimulated an active market, market prices are more than $1,000 per acre-foot ($810 per ML) (Colby, 1990b).

In contrast to the Southwest and intermountain states, few water rights have been sold in the Pacific Northwest and the northern Great Plains states. In general, water supplies far exceed demand during normal flow years in these areas, although demand for instream uses—especially to restore fisheries and support white water rafting—is increasing. Cities or businesses needing additional water can usually appropriate additional water rights under state law or can lease long-term supplies from a reservoir owner with surplus quantities.

California is the only rapidly growing arid state that has not had many water rights sales until recently (for reasons explored in Chapters 10 and 11), even though some parts of the state face significant long-term water shortages. In California, there are hundreds of privately owned surface and ground water rights that could be marketed.

---

[1] Water users and water managers in the West have a long history of measuring water in acre-feet, which is thus the primary unit used to describe water volume in this report (1 acre-foot is the volume of water required to cover 1 acre of land to a depth of 1 ft). The corresponding international system unit would typically be cubic meters (1 acre-foot equals 1,233.5 m^3). This report elects to use a less common metric unit—the megaliter (ML), where 1 ML equals 1,000 m^3—because the resulting conversions are more similar in scale to acre-feet. For instance, 100 acre-feet of water equals 123 ML (versus 123,350 m^3).

Still, much of the state's water supply is "locked in," delivered under long-term contracts with state and federal water projects. Many different parties have to agree before these contracts can be changed, so transfers of water are hard to implement. However, high demand for more water by California cities has led to water leases and innovative exchange arrangements that circumvent these difficulties.

TYPES OF WATER TRANSFER OPPORTUNITIES

Several different types of transactions—including water leases, water banks, dry year option arrangements, and transfers of salvaged water—may be used to transfer water use from one party to another. Water rights may be sold or leased, and the transfer may be permanent or temporary.

Water Leases

A water lease occurs when a water rights owner and a new user negotiate an agreement to use a fixed quantity of water over a specific period of time, instead of purchasing a permanent right. Leases often occur during dry years, when some farmers or cities run low on water supplies in storage and need a temporary way to endure short-term drought. For example, junior rights irrigators with orchards and other high-value perennial crops sometimes lease water on a one-time basis for their late summer irrigations from neighboring seasonal crop growers who hold more senior rights. Orchard crop owners are willing to pay more for this water than it was worth to irrigate field crops because they face the danger of losing their long-term investment if their trees die.

Water lease prices in various areas of the West cover a broad range. For example, in 1988 the Bureau of Reclamation offered to lease water from its Green Mountain Reservoir in western Colorado at $6 per acre-foot ($4.85 per ML) for agricultural use, $10 per acre-foot ($8.10 per ML) for municipal use, and up to $80 per acre-foot ($65 per ML) for industrial use. In the same year, the bureau's central Arizona project office leased surplus water to Phoenix-area customers at prices ranging from $35 to $82 per acre-foot ($28.40 to $66.50 per ML) (*Water Market Update*, 1988). In another lease arrangement in the late 1980s, the Montana Fish, Wildlife and Parks Department paid $20,000 for a release of 10,000 acre-feet (12,335 ML) from Painted Rocks Reservoir into the Bitterroot River to preserve downstream fisheries (Colby, 1990a).

Water Banks

Water banks are another transfer-related option. A water bank is a formal mechanism for pooling surplus water rights for rental to other users. In Idaho, for example, farmers with surplus entitlements from federal projects sell more than 100,000 acre-feet (123,400 ML) annually through water banks that are sanctioned by the state. Water bank leases generally result in changes in point of diversion of storage water or changes in place or purpose of use. Several tests must be met before reallocation through a water bank is approved, such as whether the lease would cause the use of water to be expanded beyond that authorized under the water right or whether it would conflict with the local public interest. Idaho water bank prices in the late 1980s ranged from $2.75 to $5.50 per acre-foot ($2.23 to $4.45 per ML) for one-time use of the water during the irrigation season. Part of the fees collected goes to the entity supplying the water to the rental pool; part goes to the water district to cover administrative costs. Prices are set by the water banks' governing boards and are actually well below the real market value of the water. Figure 1.1 illustrates the quantities of water banked and used for irrigation and power production in Idaho's Upper Snake River Water Bank since the bank was established in 1979.

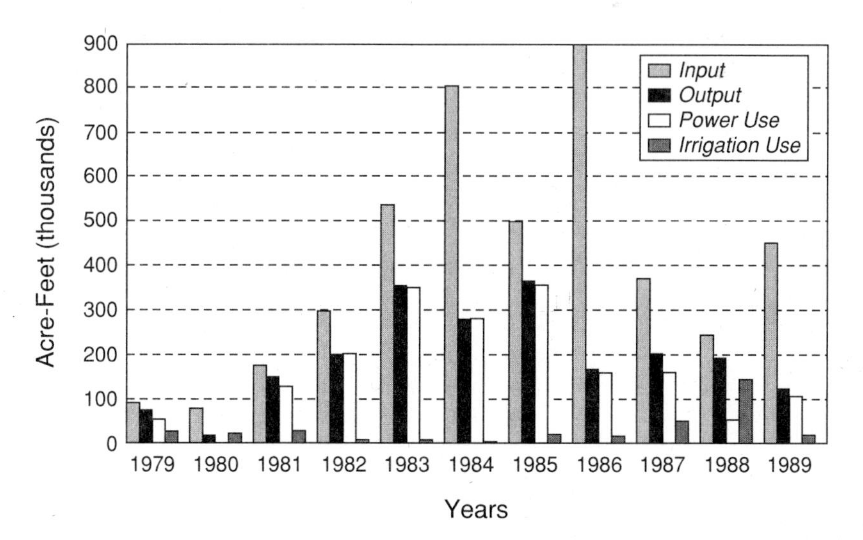

FIGURE 1.1 Summary of Idaho's Upper Snake River Water Bank activities (1979 to 1989).

In 1991, California responded to a 5-year drought by establishing a water bank to facilitate market-like transfers of water. The arrangement provides for the state to buy water from voluntary sellers and distribute it at cost to urban and agricultural users with critical needs, urgent fish and wildlife protection needs, and carryover storage to guard against a sixth dry year (Vaux, 1991).

Dry Year Option Arrangements

Many water users have enough water to meet their needs in most years but not in the driest years. As a result, users sometimes attempt to negotiate an option agreement with senior rights holders to use the senior water during dry years only. Dry year option arrangements allow the senior rights holders to continue to use the water (in most cases for farming) in normal years and give the option holder (often a municipal user) a cost-effective way to make its supply more reliable during dry years. For example, a dry year option agreement has been implemented by a Utah city and a nearby irrigator. The city paid the irrigator $25,000 for entering the option arrangement for a 25-year period; during those dry years in which the city takes water, it pays the farmer a set sum plus the quantity of hay the farmer might have grown. The farmer benefited from the cash payments and the guarantee of hay for his livestock; the city was assured more reliable supplies.

On a larger scale, the Metropolitan Water District (MWD) of southern California has proposed a dry year option arrangement to farmers in the Palo Verde Irrigation District. The MWD offered cash payments for each acre placed in the program and additional payments each time it asserts its option to transfer the water during dry years. The irrigators declined that offer, but negotiations continue. It is inevitable that more such agreements will be negotiated in the future. In northern California during the summer of 1988, the East Bay Municipal Utility District (EBMUD) offered irrigators a dry year option based on a payment for the water of $50 per acre-foot ($40.50 per ML). However, the irrigators felt the price was too low, and no agreement was reached (*Water Market Update,* 1988).

Transfers of Salvaged Water

Transfers of salvaged water also occur. This is a variation of a water sale, in which a city or business that needs additional supplies finances irrigation improvements in exchange for rights to use the water that is conserved. In Wyoming, the city of Casper paid for upgrading irrigation systems in the Alcova Irrigation District in order to salvage several thousand acre-feet of water from the district for new municipal use. In California in 1989, after years of negotiations, MWD and the Imperial Irrigation District (IID) reached an agreement calling for MWD to pay for irrigation system improvements within IID in exchange for rights to use the water conserved. The IID is at the lower end of the river system, and there are no opportunities to reuse return flows, so the return flow is considered "wasted." In California, MWD has begun a closely watched pilot program with the Coachella Valley Water District to salvage Colorado River water imported to southern California via leaky canals. Through a multimillion dollar canal-lining project, MWD hopes to salvage up to 30,000 acre-feet (37,000 ML) annually for municipal use. Additional transfer arrangements involving water conservation are under serious consideration elsewhere in California.

Other Types of Exchanges

Water transfers also include exchanges in which one user trades some water or combination of water and money for another user's supply because the timing, guarantee of availability, or quality makes that supply more attractive to the first user. This type of exchange is relatively common in Colorado, where cities buy water rights in adjacent basins and exchange them for water that can be piped through existing conveyance systems.

California droughts have spurred exploration of water exchanges. To protect the quality of supplies for urban customers during the anticipated 1989 to 1990 drought, EBMUD, in the San Francisco region, wanted to trade low-quality water to local irrigators in exchange for an equivalent amount of their entitlement to higher-quality mountain runoff from the Mokelumne River, EBMUD's normal source of supply. Again, this proposal was rejected by local irrigators.

Another bank-like type of exchange that has occurred in California and could be used elsewhere involves trading surplus surface waters in wet years for accumulated ground water supplies during droughts. The MWD of Southern California has had a policy of storing imported water in ground water aquifers since 1931 (MWD, 1989). The MWD and other water users in California are recharging and stabilizing ground water aquifers by putting surplus surface water into the ground water in both adjudicated and unadjudicated basins. This activity could be extended to irrigation districts. Surplus surface water would be stored during wet years in exchange for use of local irrigation rights during droughts. During dry years, MWD would use the farmers' surface water rights for municipal use; the farmers would then pump the water that MWD has previously recharged into the ground water beneath their lands.

THE CASE FOR TRANSFERS

The demand for water by irrigated agriculture is likely to continue its decline relative to other water uses because of a combination of market forces and a fundamental shift in values (Figure 1.2). Lower agricultural commodity prices, comparatively higher energy costs, and the rising opportunity costs of capital all hinder the development of major water projects for agriculture. Environmental concerns about dewatering and irrigation-related water quality problems are increasingly evident. Federal and state policymakers need to ensure that water allocation laws can respond to all water use demands—old and new—in an efficient and fair manner. Changes in laws and policies will be required to encourage better water management among all users.

Many environmentalists, water experts, and urban suppliers have endorsed water marketing as a desirable reallocation policy. Water marketing can help promote both efficiency and fairness. Markets respond to price signals and move resources from lower- to higher-valued uses. They also respect existing property entitlements and thus allow water rights holders both to set the pace of transition and

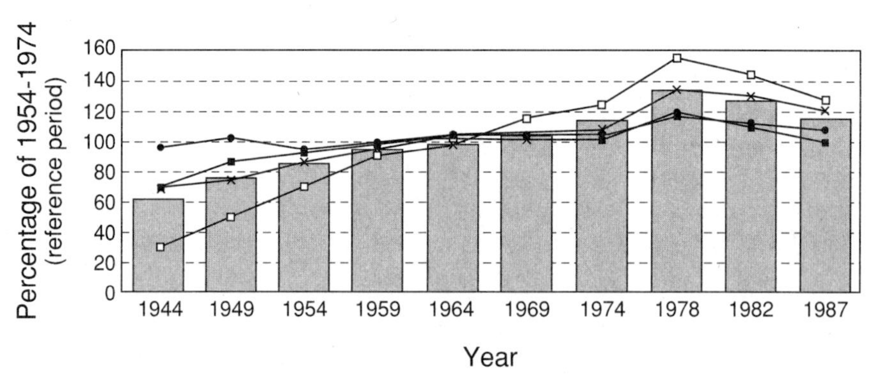

FIGURE 1.2 The area of land irrigated reached its zenith in the 1970s and is now declining in each of the regions of the West. The greatest decrease has occurred in the high plains of Texas because of declining water table levels, decreasing well yields, increasing pumping costs, and low farm commodity prices.

to receive compensation when water is transferred. For these reasons, water markets will probably be a keystone of water policy in the next century. Nevertheless, they should be seen in the context of the full range of water management techniques, for the issue is not promoting transfers per se but promoting better water management.

Federal agencies have a large and potentially positive role to play in water transfers. The Bureau of Reclamation stores and distributes large blocks of water used primarily for irrigated agriculture but that could be a source of transfers to environmental and urban uses. In basins such as the Upper Missouri River, the Army Corps of Engineers controls a great deal of water that could be the source of similar transfers. The basin case for the transfer of federally supplied water is to correct inefficiencies that are the result of a long history of subsidized irrigation water. Many students of reclamation policy have reached the conclusion articulated in a recent study (Wahl, 1989): "Rather than attempting to reduce the subsidies embodied in existing contracts, federal policymakers should seek to make the current property interests in federally supplied water more secure and to allow voluntary market trading of the resource among water users." Many federal project beneficiaries do not agree with this conclusion, but it seems clear that transfer policy will play a central role in the debate about the Bureau of Reclamation's future.

The issue is not just whether water markets can reallocate water when the gains are positive and discourage reallocation when the costs are high but whether the institutional setting is sufficient to bring all relevant third parties into the deliberations. This issue is not often raised with respect to markets because it is typically assumed that markets provide an accurate measure of the relevant values of a resource. But markets are not always efficient (Livingston and Ruttan, 1990), and they may not reflect the full range of noneconomic values.

The diverse and strongly held economic and cultural values associated with water suggest that we need an expanded set of criteria to evaluate transfers. Accordingly, in preparing this report the committee recognized the relevance of both economic techniques, which can be used to measure the value of water and the costs of transfers, and other, more subjective methods that permit subtle and intangible values to be considered.

In evaluating third party effects, the Committee on Western Water Management assumed that

- reallocation of water among uses will be a principal feature in a new era of western water management, and increased conservation, increased use efficiency, and improved reservoir operation will also be important features;

• the general direction of reallocation will be from agricultural to municipal, industrial, recreational, and environmental uses;

• water markets involving willing buyer-seller transaction opportunities will continue to expand; and

• new formal and informal constraints on water transfers and third party effects of water transfers will be established to ensure that reallocation processes include the consideration of all relevant interests.

The committee approached its examination with an optimistic sense of the role water transfers can play in a new era of more efficient use and more evenhanded response to diverse values, but with the important caveat that transfers need to be more closely monitored than they have been in the past. Judicious intervention in water transfer processes will be necessary to avoid or to ameliorate the adverse effects of some transfers. This report, like others before it (Driver, 1986; Smith, 1988), sees great potential benefits in reform of the law and administrative procedures guiding water transfer activity in the western states.

REFERENCES

Athearn, T. G. 1986. P. 35 in The Mythic West in Twentieth-Century America. Lawrence: University of Kansas Press.

Brickson, B. 1991. Water farming in the West: The impacts and implications of long-term, rural-urban ground water transfers in four western states. In Western Water. Sacramento, Calif.: Water Education Foundation (September-October).

Colby, B. G. 1990a. Enhancing instream flow values in an era of water marketing. Water Resources Research 26(6):1113-1120.

Colby, B. 1990b. Sources of water: Agriculture—The deep pool? In L. MacDonnell, ed., Moving the West's Water to New Uses: Winners and Losers. Boulder: University of Colorado, Natural Resources Law Center.

Dozier, E. P. 1970. The Pueblo Indians of North America. New York: Holt, Rinehart and Winston.

Driver, B. 1986. Western Water: Turning the System. A report to the Western Governors' Association from the Water Efficiency Task Force. Denver: Western Governor's Association.

Dunbar, R. 1983. Pp. 9-35 in Forging New Rights in Western Waters. Lincoln: University of Nebraska Press.

Gibb, H. A. R., ed. 1960. P. 492 in The Encyclopedia of Islam. Vol. 1. Leiden, Netherlands: E. J. Brill/Lozac and Company, London, England.

Leopold, L. 1990. Ethos, equity and the water resource. Environment 32(2):17-20, 37.

Leshy, J. 1990. The prior appropriation doctrine of water law: An emperor with few clothes. Journal of the West 29(3):5-13.

Livingston, M. L., and V. W. Ruttan. 1990. Efficiency and equity in institutional development: A perspective on water resources in the arid west. Rivers 1(3): 218-226.

MacDonnell, L. 1990. Shifting the uses of water in the west: An overview. In L. MacDonnell ed., Moving the West's Water to New Uses: Winners and Losers. Boulder: University of Colorado, Natural Resources Law Center.

Metropolitan Water District (MWD). 1989. Programs to Expand Groundwater Conjunctive Use in Southern California. Los Angeles, Calif.

National Academy of Sciences (NAS). 1968. Water and Choice in the Colorado Basin: An Example of Alternatives in Water Management. Washington, D.C.

Powell, J. W. 1962. Report on the Lands of the Arid Region of the United States. P. 23 in W. Stegner, ed. Cambridge, Mass.: Belknap Press of Harvard University.

Smith, R. T. 1988. Trading Water: An Economic and Legal Framework for Water Marketing. Washington, D.C.: The Council of State Policy and Planning Agencies.

Solley, W. B., C. F. Merk, and R. R. Pierce. 1988. Estimated use of water in the United States in 1985. U.S. Geological Circular 1004. 64 pp.

Stegner, W. 1987. P. 6 in The American West as Living Space. Ann Arbor: University of Michigan Press.

Vaux, H. J. 1991. The California drought: 1987-? Water Science and Technology Board Newsletter 8(2), April.

Wahl, R. W. 1989. Markets for Federal Water: Subsidies, Property Rights, and the Bureau of Reclamation. Washington, D.C.: Resources for the Future.

Water Market Update. 1988. Stephen Shupe ed., Market strategies pursued by East Bay MUD, 2(9):13. Santa Fe: Shupe and Associates. September.

Webb, W. P. 1931. Pp. 237-238 in The Great Plains. New York: Grosett & Dunlap.

Willey, Z. 1985. Economic Development and Environmental Quality in California's Water System. Berkeley: Institute of Governmental Studies, University of California.

2

Third Party Impacts and Opportunities

Sooner or later in life, we all sit down to a banquet of consequences.

Robert Louis Stevenson

A water buyer and seller are the two primary parties in a water transfer, each of whom must be satisfied with the results of the negotiations for a transfer to be consummated. These primary parties negotiate in their own best interests and exercise control over whether a transfer will occur. Consequently, their interests are not typically a central concern of public policies governing water transfers. Instead, public policies must be concerned with the interests of so-called third parties, that is, those who stand to be affected by the transfer but are not represented in the negotiations and lack control over or input into the processes by which transfer proposals are evaluated and implemented.

The impacts of transfers and the parties affected are many, diverse, and potentially substantial. Third parties are described in detail in Chapter 4; they can include

- other water rights holders;
- agriculture (including farmers and agricultural businesses in the area of origin);
- the environment (including instream flows, wetlands and other ecosystems, water quality, and other interests affected by environmental changes);
- urban interests;
- ethnic communities and Indian tribes;
- rural communities; and
- federal taxpayers.

The types of impacts felt by these parties are quite varied but can be broadly thought of as economic, social, and environmental. Economic effects include impacts on incomes, jobs, and business opportunities. Social impacts include changes in community structure, cohesiveness, and control over water resources, and such changes can occur in both rural and urban communities. Environmental effects are broad based, including effects not only on instream flow, wetlands, and fish and wildlife, but also on downstream water quality and on recreational opportunities that are dependent on streamflows, riparian habitats, and aesthetic qualities.

Because local governments in the area of origin are seldom the buyers or sellers in water transfer transactions, their interests and those of community residents frequently are of concern. Damage to the environmental and aesthetic amenities of natural and rural areas may be significant. For example, transfers that involve surface waters may decrease instream flows, leading to degradation of wetlands and water quality and to loss of riparian habitat. Such transfers also can result in increased sewage treatment costs to municipalities that rely on the depleted streams. Where surface water and ground water are closely linked, the export of ground water also can alter surface flows, with potential adverse effects on riparian vegetation and wetlands. Ground water transfers may lower the water levels in the aquifer, affecting other water users pumping from a common aquifer, drying up wetlands, and altering riparian vegetation and wildlife habitat. Negative effects tend to be most serious when transfers involve moving water from one watershed or region to another. In such instances, the benefits associated with that water are lost to the local area. Fiscal impacts include loss of property tax base and bonding capacity, tighter spending limitations, and reduced revenue sharing.

Water transfers from agricultural to other uses may lead to the retirement of irrigated land. Environmental consequences include soil erosion, blowing dust, and tumbleweeds, which arise after crop production ceases (Woodard, 1988). When farmland is retired from production, the loss of agricultural jobs and related businesses may inhibit future economic growth in the area of origin. When the tax base shrinks, causing local services to decline, the area of origin becomes less attractive to new businesses. Also, water and land resources needed for new local development may be unavailable as a result of major water transfers.

The committee considers voluntary water transfers to be the most promising approach to reallocating water, but it recognizes that voluntary transfers may be related to pressures for involuntary water allocation. Judicial and administrative rulings and water reallocations often provide a stimulus for voluntary water transfers. Chapters 5, 8, and 11, which discuss the Truckee and Carson basins in Nevada, the Yakima basin in Washington, and the Imperial Valley in California, respectively, illustrate the relationships between voluntary and involuntary transfers.

Water transfers are viewed by many as a valuable means of meeting the changing water needs of the West, and they are taking place. A recent study, for example, determined that from 1975 to 1984 some 3,853 transfer applications had been filed in Utah; 1,133 in New Mexico; 858 in Colorado; and 42 in Wyoming (MacDonnell, 1990a). The average quantity of water involved in a transfer ranged from roughly 6 acre-feet (7 megaliters (ML)) in Utah to 10 acre-feet (12 ML) in New Mexico and 11.5 acre-feet (14 ML) in Colorado. In contrast, Wyoming's few transfers tended to involve substantially larger quantities of water, with a median amount of nearly 900 acre-feet (1,110 ML).

In a 1986 survey the Western States Water Council found a wide range from state to state each year in the number of transfers (Johnson and DuMars, 1989). The differences depended primarily on the relationship between unappropriated supplies and anticipated demand. For example, in North Dakota, where unappropriated water has been available to meet new water needs and demand for new uses is low, few transfers occurred. At the other extreme, the survey found that water rights are freely bought and sold on the open market in the rapidly growing states of Colorado, Utah, Nevada, and New Mexico, where hundreds of transfers occur each year. Other studies also suggest that water transfers are increasing in number and are occurring in more areas than previously (Higginson-Barnett Consultants, 1984). These reports indicate that the higher level of transfer activity in some states reflects both the full use of available water supplies and the general level of support for transfers found in state law and procedures (MacDonnell, 1990a).

A 1990 study also found considerable differences among states in transfer approval rates, as well as differences in the number of months taken to make a decision on a transfer application and in the costs incurred by transfer applicants and objectors in the course of the state review process (Colby et al., 1990). These studies raise some concerns about the barriers posed by transaction costs but conclude that current transaction costs do not appear to be excessive in light of the need to protect the expectations of water rights holders sharing a common source of supply. As one report acknowledges, "[p]erhaps the major policy challenge facing the western states in this area is how to address third party effects" such as instream flows, recreation, area-of-origin equity, and water quality associated with the reallocation of western water (MacDonnell, 1990b).

Water Transfers in the West: A Study of Six States

Western water lawyers have always said that water flows uphill toward money. The shift in water policy from supply augmentation to the reallocation of existing supplies makes it important to know the truth behind this dictum.

Many in the water community believe that the web of transfer procedures and substantive rules imposed to protect junior water rights holders will impede market transfers of water rights. To explore this element of water transfers, the U.S. Geological Survey sponsored a major empirical and analytical comparative study of water transfers in six western states (Arizona, California, Colorado, New Mexico, Utah, and Wyoming) to determine the level of transfer activity, to identify the major legal and institutional factors that influence the efficiency and equity of these transfers, and to measure the transaction costs imposed by the transfer process. This study, *The Water Transfer Process as a Management Option for Meeting Changing Demands* (MacDonnell, 1990a), has made an important contribution to our understanding of water transfer issues.

The study concludes that existing third party protection rules do not impede transfers, although the time necessary to obtain transfer approval and the costs of this approval can be significant. Despite the imprecise nature of water rights, the study found that a great deal of transfer activity occurs throughout the West, ranging from informal trading of irrigation rights to exchanges, and from temporary transfers to permanent changes of the use and ownership of water rights, depending in part on the availability of unappropriated water in the state.

All states have removed per se restrictions against severing water from the land on which it has been historically used, but legal restrictions still exist in several states, especially on transfers by irrigation water supply organizations. Marginal adjustments in individual state procedures can be made to further facilitate transfers, but the laws as currently written and applied generally support transfers. Most but not all transfers are reviewed by a state agency, or in Colorado by a water judge, and "[t]he approval rate . . . ranged from over 94 percent in New Mexico to about 74 percent in Wyoming. Moreover, actual denials were quite rare in Colorado, New Mexico, and Utah" (MacDonnell, 1990a). The approval period and transaction costs of the transfer vary considerably among the states, but the costs of obtaining state approval for a transfer, except perhaps in Colorado, generally are not excessive. Thus the study concludes that the most pressing need is not to streamline state transfer procedures, but rather to address the major policy challenge involved in developing broader standards for evaluating third party impacts.

Indeed, arguments favoring voluntary market transfers are tempered by evidence of damage to water rights holders, adverse effects on areas from which the water is taken, impaired water quality, and reduction of instream flows for fish, wildlife, and recreation. These concerns create a dilemma for public policymakers and administrators. Market advocates urge relaxing government restrictions to facilitate markets in water as an alternative to government subsidies and regulation policies, which once were the predominant method of allocating water in the West, whereas others seek greater protection of public interest values not recognized through market mechanisms, leading to greater governmental involvement and increased restrictions on transfers (Graff, 1986; Wahl, 1989). These competing concerns demand innovative responses to balance the economic gains that transfers bring to the water reallocation process against the need to safeguard interests unprotected by the market mechanism. The desire to identify such innovative responses is a primary purpose of this report.

Following a discussion on existing legal protection for third parties in water transfers, this chapter reviews the kinds of third party impacts associated with these transfers, drawing on the committee's analysis and observations of water transfers. For each group of third parties, it describes different impacts and concerns but does not attempt to evaluate this magnitude or to review methodologies for measuring them. The chapter concludes with examples of how water transfers can be used to mitigate undesirable impacts of existing water allocations and of prior transfers. Thus, although the primary emphasis of the chapter is on the undesirable side effects that may arise as water is moved to new places and uses, the opportunities that transfers present to resolve existing conflicts and accommodate new water needs also is emphasized.

PROTECTING THIRD PARTIES

Chapter 3 examines in detail the legal framework for evaluating proposed transfers and considering potential third party impacts. In general, other water rights holders receive more protection in state review processes than any other third party. Water laws in the western states protect affected water rights holders from damage resulting from changes in use in order to make their rights more secure and valuable. Recent studies indicate, however, that water rights holders entering the process as protestants may incur substantial costs in protecting their rights (Colby et al., 1990).

The laws of some western states allow consideration of the effects

Transaction Costs: Their Role in Water Transfers

Transaction costs, in the economics literature, are the costs of making a market system work—defining property rights unambiguously enough so that sales can take place, generating information about commodities available, searching for trading partners, negotiating terms of exchange and contract provisions, and enforcing both property rights and contracts to buy and sell. In western water markets, transaction costs are incurred in searching for water supplies and for willing buyers and sellers, ascertaining the characteristics of water rights, negotiating price and other terms of transfer, and obtaining legal approval for the proposed change in water use. This latter category of transaction costs can be called policy-induced transaction costs (PITCs). Transfers participants incur PITCs as they seek to obtain state approval to transfer a water right to a new place and purpose of use. PITCs may include attorneys' fees, engineering and hydrologic studies, court costs, and fees paid to state agencies.

PITCs are of particular interest because they are the focal point of tension between two goals: the need to broaden the range of interests represented in the transfer approval process (which will increase PITCs) and the need to reduce unnecessary impediments to desirable water transfers (which implies a reduction in PITCs).

Several studies provide insights on the magnitude of costs incurred by applicants participating in state review processes. Colby et al. (1990) found that applicants' PITCs averaged $91 per acre-foot ($74 per ML) of water transferred, with considerable variation among states. Applicants' PITCs averaged $187 per acre-foot in Colorado, $54 in New Mexico, and $66 in Utah ($152, $44, and $54 per ML, respectively).

Protests were found to have a significant and positive impact on applicant costs per acre-foot. Time delay, a measure of applicants' costs of water, was also significantly related to whether or not protests were filed. In Colorado, transfers involving water rights in the most water-scarce areas of the state had significantly higher applicant unit costs and time delays than elsewhere. PITCs are higher where the economic values that may be affected by a proposed transfer are higher—in areas where water is more scarce and water rights sell for a higher price (Colby et al., 1990).

MacDonnell (1990a) also compared characteristics of water transfers in several western states and found considerable variation among states. In Colorado, 80 percent of transfer applications were eventually approved over the 10-year study period (1975 to 1984), whereas 90 percent were approved in Utah and 95 percent in New Mexico. (Approval does not imply that the applicant obtained permission to transfer as much water as originally requested; many approvals are conditioned on modifying the original transfer proposal to satisfy objectors.) One measure of PITCs involves the time delays while waiting for a state agency's decision on a transfer proposal. This period is measured in months from the time a transfer application is filed to the date of the state agency's decision. MacDonnell found that 60 percent of all transfers were protested in Colorado over the period 1975 to 1984 and that it took an average of 21 months to obtain state approval. In sharp contrast, only 5 percent of transfer applications were protested in New Mexico over this same period, and the average time to obtain approval was 5.8 months. In Utah, about 15 percent of change applications were protested, and the average time to obtain approval was 9 months.

One explanation for the variation among states in time delays, approval rates, and frequency of protests lies in the different types of transfers occurring in each state. Transfers out of agriculture often are more controversial than transfers among farmers because of effects on the area of origin and fears of economic dislocation. Transfers out of agriculture account for 80 percent of the transfers in Colorado and only 30 to 40 percent of the transfers that have occurred in New Mexico and Utah over the last 10 years (MacDonnell, 1990a).

of transfers on the special interests of areas of origin and the environment. Most states provide that transfers may be subject to a public interest review (Colby et al., 1990). Indeed, the courts in some states have involved the public trust doctrine to require such consideration even where there is no provision in state statutes. The review process can be expensive; therefore affected parties who do not have the financial means to employ attorneys, hydrologists, engineers, and other experts to substantiate potential damages are unlikely to be effective participants in a review proceeding. Federal and state environmental laws on water quality, wetlands, and endangered species protection also may restrict transfers or increase their costs by requiring that alternative water be provided for environmental needs.

In summary, although some protection exists for parties affected by water transfers, serious questions arise as to the scope of legitimate interests that should be protected, the extent of protection that should be afforded, and how it should be provided.

RURAL COMMUNITIES

Economic and Fiscal Impacts

The impact of water transfers on rural communities is an issue in several of the cases highlighted in this report. No issue gave the committee more trouble than the question of how to characterize and evaluate the effects of water transfers on small communities. The reason is obvious: no consensus exists within our society about the value of these communities. The communities generally have no legal right because we view them as inferior units of central state governments, and we generally allow the market to dictate their fate. Nonetheless, we do value many rural communities, and we are sometimes willing to buffer them against market pressures. The widespread use of historic preservation controls is one example of such a buffer. As the nation becomes more urbanized and homogenized in the twentieth century, the virtues of rural communities are being extolled with a Jeffersonian fervor. However, many of the justifications for the preservation of rural communities simply reflect an elegiac view of the past that cannot serve as the basis for contemporary public policy. This report does not attempt a comprehensive analysis of the issue of rural community preservation but does suggest processes and factors that can be considered to decide which rural communities should be protected and how this might be done.

Retiring irrigated land can lead to losses of farm jobs, crop production, and farm income. These direct effects can be measured with a fair degree of accuracy. However, indirect impacts of water transfers, such as losses of off-farm jobs, income, and production in nonfarm businesses and households, are more difficult to estimate. Another type of economic impact, "induced impacts," includes changes in population, employment, and income in local businesses and activities not linked to agriculture but dependent on the vitality of the local economy in general. Retail stores, restaurants, and local services may be affected by a decline in agriculturally linked jobs and income (Charney and Woodard, 1990).

Some economists have argued that water transfers will not impose large and sudden shocks on rural economies because only the marginal lands, those least suitable for crop production and least profitable to irrigate, will be sold and farmers will simply concentrate their efforts on their remaining high-quality lands and most profitable crops. However, there is growing evidence that it is not only the marginal agricultural acreage that is being purchased (Howe et al., 1990). Potential water buyers are willing and able to buy out the properties with the most secure and senior water rights to high-quality water sources, in the most convenient locations, regardless of the crop being grown or farm profitability. In some of the case study areas, particularly Arizona and Colorado, high-value cropland, such as pecan orchards, has been purchased, and more economically marginal grain fields have been ignored. Water buyers seek senior rights to water sources that can be conveyed easily to the new location of use, and generally whole farm operations are purchased, not just less productive portions of farms. The "marginality argument," which suggests that economic impacts on rural areas will be gradual and occur in small increments, thus appears to be largely baseless.

Although the impacts caused by transferring water formerly used in farming to other areas and uses are difficult to quantify, and may be small in relation to a state's entire economy, they are significant to area-of-origin residents. Although each individual water farm purchase may involve only a small fraction of the area of origin's total land and water resources, it is important to consider the cumulative effects of such purchases. Factors that make property attractive for water farms in La Paz County, Arizona, have led purchasers to concentrate in a few areas adjacent to aqueducts that can convey water to new uses. This clustering effect also is apparent in the Arkansas River Valley in Colorado. The result is that local economic impacts of transfers are borne by specific towns and counties.

Local economic consequences of water transfers are felt at several different times. Some occur when land and water rights are purchased; others when the land is retired from irrigation; still others when the water is actually transported to a new area. If mitigation requirements are to be effective, they must be timed to remedy impacts when they actually occur. Local government fiscal losses, for instance, may occur well before water is actually transferred out of an area, and thus some earlier event—such as the purchase of land and water rights—must trigger a mitigation requirement for fiscal impacts.

Direct fiscal impacts, including the loss of property tax base and bonding capacity and reduced debt limit and state revenue sharing, occur immediately upon purchase of the land by a municipality or other tax-exempt entity. For instance, one purchase of irrigated farms by an Arizona municipality removed 10 percent of the taxable land in La Paz County, Arizona, from the tax rolls (Nunn and Ingram, 1988). When land purchased for its water rights is removed from the county tax rolls, county tax rates must then be increased, placing a heavier burden on the remaining taxpayers, or services must be cut. At the same time, the county's bonding capacity and legal debt limit, which are based on the county's net valuation, are decreased. Net valuation is also the basis for the distribution of state-shared revenues, particularly the state-shared sales tax. This means that a county receives a progressively smaller portion of state sales tax revenue as municipally owned acreage increases.

Counties having only a small percentage of privately owned land are particularly vulnerable to the fiscal impacts of retiring irrigated land, particularly in areas where lands are acquired by a tax-exempt entity, such as a municipal government seeking new water supplies. In La Paz County, Arizona, 88 percent of the land is federal or tribal, and the state owns an additional 7 percent. This leaves less than 5 percent of county land in private ownership and subject to property taxes. In the late 1980s, 49 percent of this private land was purchased, or was under an option to be purchased, for the appurtenant water rights (Charney and Woodard, 1990).

Direct fiscal impacts also can occur when the buyer is not a tax-exempt entity. When water rights are sold and the irrigated farmland is retired, this typically results in a reclassification and reduced valuation of the land for property tax purposes.

Induced economic and fiscal impacts begin to occur after the farmland is retired. In an area where production in irrigated agriculture is reduced because of water transfers, the farms that remain may be insufficient to support some or all of the local packinghouses and seed, fertilizer, and machinery distributors. Similarly, as irrigated agriculture declines, the community becomes less prosperous, and both the economic infrastructure and the social infrastructure of the community decline. Banks, pharmacies, and other essential firms close. The social structure provided by churches, civic groups, and political organizations weakens at a time when a rural community may badly need a stable, coherent social structure to cope with economic change. The population decreases as reduced job opportunities force people to move from the rural area. If a high enough percentage of the basic industry jobs are lost, the economic viability of the community may be threatened.

One concern focuses on the "ripple" effects of water transfers on communities dependent on irrigated agriculture. Just as investments in irrigated agriculture cause "linked industries" to locate in an area, transfers of water out of agriculture can cause them to leave. The economic consequences that result from disinvestment in irrigated agriculture are multiplied as concomitant disinvestment occurs in industries that support irrigated agriculture. This raises fears that traditional agricultural enterprises and associated lifestyles may no longer be viable.

Economic development is threatened as soon as water rights are sold for use in another area. A further round of impacts is felt when the water actually leaves the area and the reality of the transfer is experienced. The fiscal condition of the rural area deteriorates as direct and indirect effects are realized. A community's ability to attract new enterprises depends on both its tax rate and its spending patterns. The quality of public services, particularly the quality of public schools, is an important consideration when people and businesses are deciding where to locate. Thus, exporting all of the water available to a piece of land not only can foreclose future development opportunities for that land, but also may limit development opportunities throughout the area.

The *perceived* water supply also is important. If the area of origin has no control over its water supply and is seen as being "dried up," this perception will impede future development. Tumbleweeds and blowing dust on retired farmlands reinforce this perception (Nunn and Ingram, 1988; Oggins and Ingram, 1990). The availability of developable land also can be impaired if water purchases lock up an area's land base or increase land prices. In many parts of the West, highways, rail lines, and aqueducts run along parallel routes; land purchased for its proximity to possible water conveyance infrastructure may also be the most desirable land for commercial or industrial development.

When assigned a dollar value, the losses suffered by areas of origin may appear insignificant in comparison with the total state economy or even with the substantial benefits of additional water supply that may accrue to the new users of the water. Such losses, however, tend to be concentrated in particular areas and can seriously impair the viability of small rural communities, which may lack the economic strength and diversity to respond to such rapid changes.

Environmental Effects of Retiring Irrigated Farmland

When land is retired from irrigated agriculture, the natural process of revegetation produces a secondary succession of plant species. Succession continues until the plant community has stabilized. Russian thistle (tumbleweed) is the characteristic species in the first phase of secondary succession on abandoned farmland in much of the Southwest. Tumbleweeds are effective in dispersing their seed, and they quickly dominate the land to the virtual exclusion of all other species.

The natural succession process varies with the climate and soil type. Farmland in relatively high rainfall areas and with coarse soils may move past the tumbleweed phase in a few years. But secondary succession is slow on fine-textured soil with low rainfall. In parts of Arizona, for example, little revegetation has occurred on farmland abandoned 20 years ago. Instead, the soil has crusted over in large barren areas where no vegetation can be established, generating nuisances of blowing dust and tumbleweeds. Some rural residents have noted that the value of nearby farmland decreases when adjacent land succumbs to disuse. The negative effects of farmland retirement on wildlife, especially small game, also have been noted (Woodard, 1988).

A significant portion of the Arkansas River Valley of Colorado has been purchased and retired from agriculture by the cities of Aurora and Colorado Springs in order to transfer the associated water rights to municipal uses. The retired farmland was neglected, and local irrigators filed suit against the cities over the environmental and nuisance problems. The court approved a negotiated resolution of the case, calling for a permanent vegetative cover to be established on the land before water is removed for use elsewhere. As a result, a revegetation project is now under way in the Arkansas River Valley, though with limited success (Weber, 1989).

Social impacts caused by water transfers, which are even more difficult to quantify, but no less genuine, include effects on community cohesion, local traditions and cultural values, and the political viability of local governments and irrigation districts in the area of origin. One pervasive effect of water transfers on areas of origin is loss of local self-determination as the future of an area moves beyond the control of its residents. It is this sense of uncertainty, frustration, and vulnerability, as much as the visible, tangible damage, that is fueling the demands for regulation of water transfers from rural areas of the West.

A Cautionary Note on Area-of-Origin Protection

Several factors are relevant to evaluation of the indirect effects transfers can have on rural economics, and these need to be considered before a policy of area-of-origin protection is adopted. First, to the extent that transfers of water are voluntary, the value of the water *to the buyer* in the new activity exceeds the value of water *to the selling farmer*. If this were not so, the voluntary sale of the water would not have taken place. Whether this transfer constitutes a net economic gain depends on the area or region in which benefits and costs of the transfer are being compared. Evaluation of the net economic effects of transfers out of agriculture depends on the size of the area in which benefits and costs are considered. A transfer can appear to be negative from a county-of-origin perspective but result in positive net benefits when viewed from a statewide perspective. For instance, one study of the Arkansas River Valley in Colorado found that recent transfers resulted in a net income loss of $53 per acre-foot ($43 per ML) but that the market value of water in the urban areas exceeded $1,000 per acre-foot ($810 per ML) (Howe et al., 1990).

Although water transfers can bring negative effects, it is important to recognize that a dynamic, growing economy depends on processes that allow declining industries and firms to be displaced by growing firms and industries. Recent U.S. experience with the decline of several primary industries in the rust belt is an example. Government does not typically move to protect declining industries or to provide full indemnification to people displaced by industrial decline. Indeed, some government activity itself causes adverse indirect impacts for which compensation is not offered. The case of new freeways that bypass small towns and thereby diminish their economic vitality is but one example. If the economy of the West is to remain vibrant and if national and world demands for food and fiber produced in the West do not grow substantially in coming decades, then some disinvestment in irrigated agriculture is probably inevitable and desirable. Efforts to forestall rather than to effect an orderly transition are likely to be counterproductive in the long run.

AVAILABILITY OF WATER FOR AGRICULTURE

Many farmers fear that once water is transferred out of agriculture, it can never be recaptured. Some believe that the transfer of significant quantities of water to the urban sector would prevent farmers from reacquiring adequate water supplies in the future should irrigated agriculture become significantly more profitable than it is today. This concern arises even when leases instead of sales are contemplated, partially because of a distrust of the contracting process. Many growers believe that courts, state engineers, or boards would be reluctant to displace water from a "higher beneficial use" in the future simply because a lease calls for the water to return to the agricultural sector.

Well-functioning water markets tend to drive up the price of water, benefiting existing agricultural water users who want to sell or lease. But growers who want to purchase more water for farming in the future perceive correctly that higher prices will be a disadvantage to them, and so they oppose transfers in concept.

ETHNIC COMMUNITIES

Water transfers can affect the cultural integrity of communities such as the Hispanic towns of northern New Mexico and Colorado's San Luis Valley. Although Indian and Hispanic communities present the strongest case for protection, they are not the only deserving water-related communities. There are three primary reasons for their special status. First, these communities are rooted in a specific place and resource base more so than most other communities. Second, these communities have often been deprived of resources in the past. Third, members of these communities are subject to a greater degree of risk if they are eroded. Because they generally lack the political resources to prevent or to rectify these effects, such communities may be unable to survive and their way of life may be jeopardized if water has been bargained away for outside uses. Displaced community members who relocate face unique challenges in adapting to new locations compared to others displaced by changing technology and market preferences.

In many cases, there are strong bonds between water and the cultural values of minority communities. Thus the loss of control over water resources is more than just the loss of control over a resource as a commodity of monetary value; it is also an infringement on all cultural values associated with the water itself (Brown and Ingram, 1987). For instance, as explained in Chapter 7, in northern New Mexico modern water law has been imposed on a centuries-old water management approach (the community-based acequias) to the detriment of acequia integrity and viability. The prior appropriation doctrine is inconsistent with the traditions of the acequias, which hold that communities share both water shortages and surpluses. Thus imposition of the prior appropriation approach threatens the existence of these older approaches.

Loss of local control over water resources also diminishes the opportunity of people to participate in water management, which is a strong unifying bond for many communities. Irrigated agriculture is still seen by many minority communities as offering good economic opportunities and at the same time being culturally compatible with their way of life. The loss of control over water resources, however, forecloses lifestyle choices and economic opportunities that depend on local control over water.

There is some potential legal protection for ethnic communities, such as state public interest criteria. Poor rural communities often do not have the financial resources to gain access to effective legal representation in court and administrative processes. Furthermore, such statutes rarely have been used to protect their interests. Consequently, minority groups sometimes view the law as inaccessible or weighted against them and may not participate in legal proceedings even when their interests are at stake.

TRIBES AS SOVEREIGN GOVERNMENTS

Tribal governments have a distinct advantage over non-Indian rural communities in that they own extensive water rights and also have broad legal and political control of their reservation resources. Their sovereignty enables them to prevent disadvantageous transfers and to propose and shape favorable transfers. Thus they can protect their own cultural integrity. However, this authority can be exercised effectively only if they have access to sufficient information and technical expertise to evaluate proposed transactions (Williams, 1990). The Truckee-Carson case study in Chapter 5 is a good example of how a tribal government (the Pyramid Lake Piaute Tribe) influenced basinwide water management.

Even with these governmental powers, there can be significant detrimental effects on tribal communities from transfers among non-Indians. In principle, the rights of Indian tribes are not affected by water transfers or uses by non-Indians. Indian rights generally have early priority dates, and their seniority can theoretically be asserted to defeat existing uses. But, as a practical and political matter, this may be difficult or impossible. Major public and private investments have built up non-Indian dependence on water to which tribes legally are entitled but which they may not be using owing to lack of development activities. Under these circumstances, it becomes extremely difficult for the tribes to put that water to specific uses later in tribal development projects. For instance, some observers speculate that the dependence of southern California and Arizona on much of the water to which the tribes located in these areas are entitled will render it difficult for that water to be reclaimed by the tribes for their own uses. Not only will they have difficulty getting government financing for water development and irrigation projects, but there may also be political pressure against making water supplies available for use on reservations by those who have been permitted to take water in default of Indian uses.

ECOSYSTEMS

The integrity of ecosystems depends on healthy wetlands, riparian areas, estuaries, and associated fish, wildlife, and vegetation. Transfers

Leasing Tribal Water for Non-Indian Use

Off-reservation transfers of tribal water are controversial. On the one hand, many tribes feel that the right to implement such transfers is essential to reap economic benefits from their water rights, in the absence of federal largesse for capital-intensive agricultural development on the reservation. Federal funds that were available for decades to support non-Indian irrigation project development during the reclamation era were not, and will not be, available on a comparable basis to tribes. Tribes therefore may look to water leasing as a means to earn capital needed to stimulate reservation economies and alleviate poverty and unemployment.

On the other hand, state water rights holders have become the junior rights holders in many areas where senior rights have been awarded to tribes. If tribes transfer senior rights to on- or off-reservation uses other than the existing junior rights holders, the water supplies for non-Indian juniors could be disrupted. If tribes may not engage in such transfers, disruption of junior rights merely will be delayed until tribes are financially able to develop irrigation projects and other on-reservation consumptive uses.

Tribal administration of water use transfers will be challenging, given the many preexisting junior users. However, tribes understandably desire some flexibility to benefit from their water rights under changing economic conditions. Although tribal agricultural projects and other on-reservation uses can be an important component of reservation economic development, these projects require capital. Given the high economic value of water in some off-reservation uses, water use transfers are an important means to raise funds for development projects and to achieve the highest and best use of regional water supplies. Policies that limit tribes to on-reservation water uses preclude the use of tribal water to supply growing demands off reservations and deprive both tribes and non-Indians of the benefits of voluntary transfers.

of surface or ground water can have significant impacts on water-dependent flora and fauna within western riverine, riparian, and wetland ecosystems. In the arid West, these ecosystems typically occur in narrow bands along river corridors. Water development over the last century destroyed many of the large wetland complexes located at the terminus of closed-basin river systems. Entire species of plants and animals in the arid West are now threatened with extinction because of the reduction of critical river corridor habitat. At the same time, some irrigation activity actually created small wetlands that rely on the numerous leaks and seeps along earthen conveyance systems. Large riparian trees such as cottonwood and willow also came to grow along some regulated river corridors and conveyance canals. Over time, birds and mammals have become dependent on the wetlands maintained by irrigation return flow. As water is transferred, both the quantity and the quality of the water delivered to these wetlands are likely to diminish.

Several species of large river-inhabiting fish are listed as endangered because their habitat has been reduced and their migratory routes have been blocked by dams and diversions. Examples include the federally listed cui-ui in the Truckee River and the threatened salmon runs in northern California rivers and in the Northwest.

INSTREAM FLOWS AND RELATED BENEFITS

In most states, impacts on streamflows are not routinely considered when a water transfer proposal is evaluated. Thus foregone instream benefits can be a significant third party impact caused by water transfers. As noted earlier, instream flows are vital in preserving fish and wildlife habitat in arid regions. But instream flows are also critical to water-dependent recreation, and these leisure activities draw visitors and tourism dollars to the region. There are essentially four kinds of economic benefits generated by instream flows, each of which can be affected by water transfers (Colby, 1990).

1. Streamflows benefit recreationists directly. Outdoor recreation in the West concentrates around lakes, rivers, and streams. Adequate streamflows are necessary for boating (especially white water rafting), hunting, bird watching, and other wildlife-related recreation. Studies indicate a significant economic payoff in augmenting streamflows in low-flow years for recreation, even though augmentation reduces water availability for other uses (Daubert and Young, 1979).

2. Recreation-related spending contributes to local economies. Dollars spent on boating, fishing, hunting, camping, and equipment support government agencies and recreation-related businesses and stimulate local and tribal economies. Many small towns and Indian reservations rely on water-dependent seasonal tourism as a significant source of livelihood for local residents. Many studies address the local economic benefits of streamflows (Boyle and Bishop, 1984; Cordell et al., 1990; Crandall and Colby, 1991; Stevens and Rosen, 1985).

How society values water left in waterways has changed over time. Instream flows are now seen as important for water quality protection, fish and wildlife management, and habitat preservation. Instream flows also are critical to water-dependent recreation, like white water rafting, a significant contributor to many western economies. CREDIT: Friends of the River.

3. Instream flows generate "nonuser" values. Individuals who do not directly use streamflows for recreation also benefit from instream flows. Preserving a riparian area gives one the option to enjoy it in the future. Choices are often made between an irreversible alternative (or one that is very costly to reverse)—such as drying up a stream environment or flooding a canyon—and the alternative of leaving the stream environment in its current state. The latter alternative is always reversible, because new diversions of water and development projects can be approved later. "Option value" is the benefit of *not* pursuing the irreversible alternative and thus leaving resource use options more flexible for the future. Willingness to pay for preservation so that one's heirs can enjoy the preserved species or ecosystem is termed "bequest value." Benefits generated by simply

Why "Recreation" and "Fish and Wildlife" Need Separate Attention

Water transfers can affect different values in different ways, so the processes used to evaluate proposed transfers must be able to distinguish these different values. When examining the third party impacts of water transfers, the broad term "environment" often is used to encompass many elements, including ecosystem effects, recreation use, fish and wildlife management, and habitat preservation. But combining all environmentally related values into one category dulls the importance of these diverse values. At times, it even lumps conflicting values together. It is imperative to separate "environment" into its varied components to properly analyze the effects of water transfers and to guard against the tendency to focus on one or two proxies for environmental quality.

The components of the natural system affected by water transfers fall generally into two groups—the functions of ecosystems and some human use of these systems. The terms environment or habitat are most accurately used to mean the actual ecosystems that are influenced by a water transfer. These include primarily aquatic and riparian systems, but also may include related uplands.

Water for management of fish and wildlife habitat is another separate environmental interest in any water transfer. Fish and wildlife management implies a human use by certain components of the natural system. These uses, however, are managed, often for economic gains, and the relative amount of hydrological degradation or improvement created by a transfer often is judged against an existing controlled environment rather than a natural biologically diverse system.

Ecosystem preservation is related to aesthetics and wildlife habitat maintenance, and these in turn may influence the recreational value of the system. Recreation is more than a human response to an aesthetically pleasing or healthy ecosystem. Recreational use of river systems often is driven by economics. Boating, white water rafting, fishing, and waterfowl hunting are some of the recreational uses that can be directly influenced by water transfers.

Although it is often convenient to treat all environmentally related impacts under one category, the potential impacts of water transfers on recreational use should be acknowledged as separate from the impacts on fish and wildlife habitats and other ecological concerns.

knowing a unique site will continue to exist are termed "existence values."

Nonuser benefits are relevant in valuing instream flows where there are wildlife species whose survival depends on streamflows and also where there are areas whose aesthetic and recreational characteristics depend on free-flowing water. Although nonuser values are difficult to quantify, studies indicate that they are sizable, especially for unique recreational sites and for endangered species. For example, existence, bequest, and option values ranging from $40 to $80 per year per nonuser household have been documented for stream systems in Wyoming, Colorado, and Alaska (Greenley et al., 1989). Although these nonuser values may seem intangible and do not go directly into some individual's or business's pockets, as recreation expenditures do, they are an important economic consideration. Evidence on nonuser values has been important in recent court rulings and settlements in determining damages from oil spills and other environmental disasters.

4. Streamflows provide water quality benefits. Streamflow levels affect dissolved oxygen levels and other water quality parameters. As streamflows become depleted, water quality standards are more likely to be violated, and municipal sewage treatment plants and industrial dischargers have to incur additional expenses to ensure compliance with national and state standards. A stream's waste assimilation capacity provides economic benefits related to the cost of treatment that would otherwise be incurred by dischargers and by downstream water users (Young and Gray, 1972). These benefits can be measured and compared to the value of water in off-stream uses such as irrigation. In some cases, instream values can be equal to or greater than water values in consumptive uses, especially in important recreation and wildlife areas (Colby, 1990b).

Failure to consider the whole array of instream and environmental consequences may make a transfer proposal seem much more attractive than it would be if all costs were accounted for. This can result in transfers that actually reduce the economic benefits generated by regional water resources. Although there are inherent difficulties in more fully accounting for these types of transfer impacts, they will prove to be critical as water transfer institutions and public participation processes evolve.

The negative impacts of actions that deplete instream flows have been dramatically demonstrated at Mono Lake, the second largest lake in California. Mono Lake—one of the relatively few, large deep saline lakes in the world—is fed by five freshwater streams. In 1941 the city of Los Angeles acquired rights to divert water from these

Quantifying Instream Flow

Instream flow refers to that volume of streamflow necessary to maintain a specific instream use, or group of uses, at an acceptable level. Instream flow requirements are tailored to accommodate differences in the hydrologic and morphologic characteristics of individual stream segments. The most desirable instream flow requirement simultaneously satisfies several instream uses and still provides water for off-stream uses.

Planners can identify a base streamflow necessary to maintain fish habitat or forecast the response of fish habitat to naturally occurring or project-induced changes in streamflow, stream temperature, sediment transport, or water chemistry. Many of these methods were developed to resolve conflicts resulting from excessive withdrawal of water for off-stream uses. They began evolving in the 1950s and 1960s, during the most active period of construction of large federally funded irrigation and hydropower projects.

The term "minimum flow" was coined to imply protection of a low-flow level that would either constrain reservoir storage during the spring runoff period or limit irrigation diversion during the low-flow season so that adequate water remained instream. Minimum flows can be reserved to maintain fish populations or levels desired for other uses, such as boating.

Instream flow recommendations should be developed whenever a proposed project or water allocation scenario will alter streamflow, stream temperature, channel structure, or water chemistry, especially when economically significant fisheries or recreational resources are involved. The most essential step in conducting an instream flow study is identifying the purpose of the streamflow assessment—in other words, deciding what specific resources are of concern. The three major purposes for conducting an instream flow investigation are as follows:

1. to determine the minimum streamflow to be reserved during water allocation proceedings and for water quality enforcement;
2. to determine seasonal streamflow requirements necessary to maintain, restore, or enhance a particular instream use; and
3. to evaluate the effects of proposed or existing water developments and water transfers and to identify opportunities for mitigation, restoration, or enhancement of affected instream and riparian resources.

Research is being pursued to quantify instream flow requirements for protecting riparian vegetation and wetlands dependent on streams and water conveyance systems. The quantification of flows needed for rafting and kayaking is becoming an important consideration in operating reservoirs and scheduling releases on western rivers. However, quantifying and administering instream flows are complex and in reality involve far more than simply determining minimum flows. Rather, managers must understand the flow regime, which is more complicated because it involves rates of change in flow and the different timing needs of different uses of the water.

streams and, eventually, to take virtually the entire flow of the lake's feeder streams. The harmful environmental consequences were known, but it was "the established policy of the state that the use of water for domestic purposes is the highest use of water" (Casey, 1984). Since 1941, the level of Mono Lake has dropped 46 ft (15.3 m), the volume of the lake has been cut in half, and its surface area has been reduced by one-third. Mono Lake's ecological balance has been seriously compromised, and wildfowl populations have been decimated (Botkin et al., 1988; Casey, 1984; NRC, 1987). In recent years, environmental groups have used the public trust doctrine and California fish and game laws to curtail Los Angeles' diversions and to stabilize the lake level.

One study estimated total visitor and nonvisitor benefits from the preservation of Mono Lake levels to be about $40 per California household, well above the cost of 22¢ per household that would be needed to preserve lake levels by replacing the Los Angeles diversions with water from other sources (Loomis, 1987). These figures suggest that the benefits of preservation can significantly outweigh the costs of preservation and have clear implications for California and Los Angeles area water policy decisions.

Contingent valuation techniques can be useful for quantifying environmental values such as resource protection, although anyone using such numbers must be aware of the severe limitations and uncertainties of this method. Attention to the benefits generated by instream flows in the West will help to identify hidden costs in proposed water transfers. State reviews of proposed transfers may continue to favor off-stream water uses if decisionmakers rely on more easily documented economic benefits provided by water for irrigation, energy development, and urban growth. Recent evidence on the economic value of instream water suggests that instream benefits can often exceed the benefits generated by off-stream uses and that economic development in the western states could be enhanced by protecting selected stream segments for recreation, wildlife, and improved water quality (Colby, 1990b).

WATER QUALITY

Typically, state water rights proceedings to review proposed transfers do not deal with effects on water quality because water rights are based on quantity only; water quality control is maintained through separate legal frameworks. Yet water use—where, when, and how water is diverted and applied—can have a profound effect on water quality (Getches et al., 1991). For example, in the San Joaquin Valley in California, inadequate drainage of irrigation return flows led to the accumulation of toxic levels of selenium and the subsequent contamination and closure of Kesterson National Wildlife Refuge (NRC, 1989). Irrigation from the Colorado River and its tributaries affects not only the amount of water flowing downstream, but also its salt content. The Colorado River Compact and a treaty with Mexico addressed water volumes; when quality issues became severe, separate legislation was passed to address salinity. The negative impact of irrigation in Colorado on the quality of water available to the Metropolitan Water District of Southern California was recognized physically but not considered an infringement of water rights. Only after extensive negotiations were the water quality implications of U.S. use of the Colorado River incorporated into international agreements.

The economic importance of the water quality enhancement and assimilative capacity provided by streams is likely to increase as water providers and wastewater treatment facilities face more stringent water quality standards. Few studies have estimated the monetary benefits of assimilative capacity. However, Young and Gray (1972) did estimate values for assimilating biological oxygen demand. Other studies indicate substantial benefits to recreationists from maintaining and improving surface water quality in Colorado's South Platte River basin along the populated Front Range (Greenley et al., 1989).

Water quality has a significant impact on the economic value of water rights because it affects the range of different uses to which water can be put and the cost of treating water to provide a quality level suitable for specific uses. In Colorado's lower Arkansas River basin, for example, irrigation companies have had difficulty marketing water for municipal use because of water quality problems. Fort

Lyons Canal Company and Amity Canal Irrigation Company water is viewed as "unsalable" to cities because contaminants in the water must be removed by reverse osmosis, an expensive process, after transporting water long distances from its current point of use.

Once accustomed to water of a certain quality, municipalities generally consider only water of equal or better quality when searching for new and attractive sources. They pay a premium to ensure that the quality of that source is protected or to secure access to other high-quality sources. For instance, the East Bay Municipal Utility District (EBMUD), which serves more than one million customers in the San Francisco Bay area, relies on the Mokelumne River basin for nearly all of its water supplies. During critically dry years the district has had to implement water rationing and search for additional supplies. The neighboring Contra Costa Water District (CCWD) offered EBMUD use of some of its water from the Sacramento River/San Francisco Bay delta. In spite of its critical shortage, EBMUD preferred not to use this alternative source because it is of significantly lower quality than EBMUD's usual source and its treatment would have stretched the capability of EBMUD's system, which was designed only for high-quality sources. Since the early 1970s, EBMUD also has sought to transfer higher-quality water from the American River, consistent with its goal of seeking the highest-quality water available.

An economic methodology known as the "damage avoided approach" can be applied to estimate the economic value of water sources of differing quality levels. For instance, industries that use water high in total dissolved solids typically face higher operation and maintenance costs and a shorter lifetime for the equipment used in processing. The value of better-quality water can be estimated by evaluating the costs that are avoided by switching from a lower-quality to a higher-quality water source (Gibbons, 1986).

Analysis of market transactions involving water sources of differing quality levels also helps to measure the economic value of high-quality water supplies (Colby, 1990b). The cost of treating different water sources to provide a quality level required for a particular use is also evidence of the economic value of clean water.

Water quality does enter into water allocation decisions but generally through the back door. For historical reasons the right is vested in water quantities or flow rates, with quality considerations addressed through regulatory regimes. In a legal and economic context, it would be more rational if quantity and quality attributes were equal partners in the determination of a water right and also in the determination of third party injury when a transfer proposal is being evaluated.

URBAN INTERESTS

Urban population growth and municipal interests' desires for more water and more drought-proof water supplies are a driving force behind the water transfers discussed in the Arizona, Colorado, New Mexico, Nevada, and California case studies. Urban interests are often the buyers of water and therefore are often primary parties in transfer negotiations. However, urban interests are also sometimes affected by water transfers and by transfer regulations.

A significant proportion of urban water system costs is related to water quality and treatment facilities. To the extent that a transfer affects the quality of a water source for city supplies, an urban area could experience third party effects. This concern emerged in the Colorado case study in instances where cities objected to proposed transfers because they feared that the transfer would affect the quantity and quality of flows for municipal treatment facilities that were designed to cope with a particular quality of water.

Some argue that transfers may promote urban growth because water transfers are sometimes used to provide water service for new developments. During the California drought, some urban residents expressed bitterness over being required to take "30-second showers" and give up their lawns while new houses were being built "down the street," bringing more people, traffic, air pollution, and so on. Reflecting this sentiment, a popular bumper sticker in San Diego in 1991 read "Stop Growth—Flush Twice." In fact, however, the evidence seems to suggest that limiting water supplies may not be an effective means of controlling growth (Erlenkotter et al., 1979).

Policies that make it difficult or more expensive to transfer water from an existing user, such as a farmer, to a city can affect the cost of water service for urban residents. If agencies or courts must consider a broader array of third party impacts, the costs to cities of acquiring new water supplies will probably rise and so will household and business water rates. Because of the high value placed on municipal water use, it is unlikely that the additional transaction costs will render many transactions infeasible. Some of the higher costs may be passed on to consumers, but this could actually have positive effects. Studies of urban water demand indicate that city water users respond to rising water rates by cutting back somewhat on water use, especially outdoor uses for landscape maintenance, car washing, and so on. Better water conservation by city dwellers is generally desirable. In the Arizona and California case studies, rural residents voicing concerns about water transfers expressed a view that cities ought to be made to manage more effectively the supplies they already have before they are allowed to buy water away from rural areas. This issue was illustrated most dramatically in the Truckee-Carson case study, where the Pyramid Lake Piaute Tribe insisted that urban areas undertake residential water metering and water conservation as a condition for settling a decades-old dispute over allocation of water. The tribe successfully argued that these conservation measures were reasonable if urban areas expected the tribe to compromise on its own water use priorities.

FEDERAL TAXPAYERS

Federal taxpayers historically have borne the costs of federal subsidies for western irrigation development and crop production. Presumably, consumers throughout the nation have reaped some benefits in terms of lower food prices and western regional economic development, although it is doubtful that the national ledger is balanced. The era of subsidized water development is ending, but the federal government continues to be involved in western water management and particularly in resolving conflicts over reserved rights and endangered species debates. Federal cost sharing continues to be an important strategy contemplated by local interests trying to resolve tribal claims and other water conflicts. Federal financial participation is an important component of the settlement negotiated in the Truckee and Carson basins and of several recent tribal water settlements in Arizona. Thus federal taxpayers are an affected party in many water transfers.

Taxpayers are affected whenever the federal government participates financially in a water transfer or in resolving conflicts over water claims and whenever federal agency time and expertise are expended. In fact, there was no case study examined by this committee in which federal involvement was absent or peripheral to the issues. Taxpayer dollars are well spent when federal involvement promotes efficient and equitable allocation or helps balance multiple social goals. However, when federal agencies compete against one another, operate only to protect traditional bureaucratic mandates, or fail to facilitate conflict resolution and balanced pursuit of broader social goals, taxpayers' "returns on investment" may be low or negative. Some of the broad social goals that often conflict with one another include promoting low-cost food and fiber production (water subsidies and commodity programs for agriculture), resolving reserved rights claims, protecting water quality and wetlands, and preserving endangered species.

OPPORTUNITIES

Transfers should be approached not just as the possible cause of third party effects but also as a potential opportunity to resolve problems with current water allocations and prior transfers. Transfers can be designed to promote not only efficiency but also a broad range of societal interests. If these broader interests are represented, taken seriously, and addressed, there need not be "losers" associated with water transfers. There are many examples of transfers that accommodate these broader interests.

The Metropolitan Water District (MWD) of Southern California's announcement of a major water conservation and marketing agreement with the Imperial Irrigation District (IID) has been hailed as a "win-win" transfer. Under the agreement, MWD agreed to finance improvements in irrigation system water efficiency in return for transfer of the saved water to MWD. The conservation package included the construction of new reservoirs to regulate water flows within Imperial's irrigation system, automated control structures, and the lining of earthen irrigation canals to reduce seepage into the subsurface soils and the Salton Sea. Thus at the same time that irrigation efficiency is improved in the Imperial Valley, urban water supplies will be supplemented for a growing urban population that totals nearly 15 million. (Even this generally beneficial transfer has potential environmental effects that must be weighed, however, such as declining water levels in the Salton Sea.)

In 1979 the Idaho legislature created a water bank to facilitate the temporary transfer of water rights to other water users in a system with a long-term surplus. This approach actually began in the 1930s, when Idaho farmers began "depositing" water allocated to them from federal reservoirs in the upper Snake River system, to be "withdrawn" by other farmers who needed the water. These deposits and withdrawals were made on a yearly basis under a lease agreement. The water bank created in 1979 to formalize this activity is operated by the Idaho Water Resources Board, which can appoint local committees to oversee the rental of stored water (Johnson and DuMars, 1989). The principal deposits to the bank have come from farmers in the upper Snake River basin with entitlement to Bureau of Reclamation water; the Idaho Power Company has made withdrawals to generate electricity. Both parties benefit, since the farmers are paid for water that they do not currently need and the power company obtains water at reasonable rates to produce electricity, thus saving money for its rate payers.

Water transfers can help preserve and enhance public values. For example, The Nature Conservancy, an international membership organization dedicated to preserving natural diversity, has been acquiring water rights in the West to complement its long-standing programs of purchasing biologically important lands. The Nature Conservancy has developed a number of strategies to acquire water rights to protect instream flows and wetlands. One of these strategies consists of acquiring existing water rights and changing them to instream uses. In so doing, it attempts to acquire generally senior water rights through purchase or donation and then to transfer such senior water rights to instream use. The Nature Conservancy anticipates that this strategy can make a significant contribution to the protection of western instream flows (Wigington, 1990).

One example of this approach to transfers occurred on Colorado's Gunnison River. Under Colorado law, the Water Conservation Board is authorized to acquire and change existing water rights to instream use and to negotiate contractual enforcement remedies with the private parties that offer such water rights to the board. Pursuant to this authority, The Nature Conservancy reached an agreement with the board to donate a significant water right with a 1965 priority date to the board for change to instream use in the Black Canyon of the Gunnison River. The agreement specifies how the instream water right will be enforced against some large junior water rights to divert water out of the river at the Gunnison tunnel, just upstream from the Black Canyon, and gives The Nature Conservancy a contractual remedy should the board fail to enforce or defend the instream water right in general. The instream flow therefore is protected according to priorities established by the appropriation doctrine and is enforceable by both a state agency and a private entity. At the same time, other water rights holders are not adversely affected by the conversion of offstream consumptive uses to instream flows, with the resulting benefits to the recreation and fish and wildlife values of the Gunnison River.

The Nature Conservancy, the Environmental Defense Fund, and other environmental interests are also involved in acquiring existing water rights for wetland uses as part of the solution to the complex problems in the Truckee-Carson case study in Nevada. At the direction of Congress, the federal government also has appropriated money to buy existing water rights that will in turn be dedicated to restoring Stillwater National Wildlife Refuge. Many uncertainties surround the transfer of the Newlands Irrigation Project water rights to wetlands within the refuge because of the many existing water claims and the significant amount of water needed to restore the refuge. Nevertheless, it is clear that any solution that attempts to accommodate the new uses represented by the refuge and that protects existing uses, including Indian water rights, will rely heavily on market transfers (Tarlock, 1990). These transfers provide opportunities to change past water allocations that did not consider fish and wildlife needs.

A water rights transfer during the energy boom of the 1970s illustrates that rural economies can be protected as water transfers occur. The Intermountain Power Project (IPP) in Utah was designed as a 3,000-MW coal-fired power plant. A site near Delta, a town of 5,000 in west central Utah on the edge of the Great Basin, was selected with the knowledge that water rights would have to be acquired from local irrigators (Clinton, 1990). This longtime agricultural community and the surrounding area are supplied from both ground water and the regulated flow of the Sevier River. Historically, the area used about 150,000 acre-feet (185,000 ML) of water per year for irrigation. The power plant was expected to need about 45,000 acre-feet (56,000 ML) of water per year.

Water rights purchase negotiations began in early 1978. Although individual stockholders of the irrigation companies could have sold water rights to IPP, local leaders organized the stockholders into a unit for negotiating purposes. Although the market price for shares in the ditch companies historically had been in the $300 to $500 per acre-foot ($240 to $400 per ML) range, when the overall negotiations were completed in 1979, IPP paid about $1,850 per acre-foot ($1,500 per ML) of entitlement (Saliba and Bush, 1987).

Because only two units of the power plant are on line, using about 16,000 acre-feet (19,700 ML) per year, approximately two-thirds of the purchased rights of about 45,000 acre-feet (56,000 ML) per year is being leased back for local agricultural use. On average, each farmer received between $100,000 and $150,000, for a total infusion into the local economy of some $80 million. This infusion of capital into the area has had many results. Some farmers used the money to reduce debt; other purchased homes, additional supplies, or equipment; a few sold out and retired. By and large, the capital resources from the water purchases have remained in the local area and served to boost the local agricultural economy. The economy also has been the beneficiary of the jobs and tax base produced by the power plant (Clinton, 1990). The IPP transfer differs from many of the case study transfers that moved water out of agriculture because the water use remained in the area of origin and was leased to farmers, thus preventing sudden reductions in irrigated acreage. These examples indicate the potential of water transfers to address third party impacts and make water available for new uses. They demonstrate that strategies can be implemented to protect affected water rights holders and public interest values that traditionally have not been recognized in market mechanisms.

Money is not always the most appropriate or useful way to characterize the full range of social, political, and environmental effects of water transfers. Interestingly, the courts have ruled in some instances that money cannot be an adequate substitute for "wet water." For instance, in the dispute between Texas and New Mexico over interstate compact violations on the Pecos River, the court ruled that past violations could be compensated by monetary payments to Texas but that any future violations by New Mexico would have to be paid with water, not money. As another example, tribes are sometimes offered monetary payments as a substitute for wet water delivered for use on reservations, so that non-Indians can continue to use water that would be awarded to tribes. In some cases, tribes have refused, arguing that money is no substitute for water supplies (Sly, 1988).

Policy changes that afford broader protection to third parties generate important benefits by limiting the uncompensated costs that transfers can impose on third parties. However, broader third party protection also means that the transfer review process will likely become more complex and cumbersome, raising the transaction costs incurred by transfer proponents. Recent court rulings and legislative activity in the western states, reviewed in Chapter 3, suggest that policymakers are broadening the protection available, reflecting a growing appreciation of the environmental, recreational, and cultural benefits that water resources provide.

REFERENCES

Botkin, D., W. S. Broecker, L. G. Everett, J. Shapiro, and J. A. Wiens. 1988. The Future of Mono Lake—Report of the Community and Organization Research Institute Blue Ribbon Panel for the Legislature of the State of California. Davis: University of California, California Water Resources Center, Report No. 68.

Boyle, K. J., and R. C. Bishop. 1984. Lower Wisconsin River Recreation: Economic Impacts and Scenic Values. University of Wisconsin Agricultural Economics Staff Paper Series, No. 216. Madison: University of Wisconsin, Department of Agricultural Economics.

Brown, L., and H. Ingram. 1987. Water and Poverty in the Southwest. Tucson: University of Arizona Press.

Casey, E. 1984. Water law—Public trust doctrine. Natural Resources Journal 24(3):809-810.

Charney, A. H., and G. C. Woodard. 1990. Socioeconomic impacts of water farming on rural areas of origin in Arizona. American Journal of Agricultural Economics 72(5):1193-1199.

Clinton, M. J. 1990. Water transfers: Can they protect and enhance rural economies? Moving the West's Water to New Uses: Winners and Losers. Boulder: University of Colorado School of Law, Natural Resources Law Center.

Colby, B. G. 1990. Transactions costs and efficiency in western water allocation. American Journal Agricultural Economics 72(5):1184-1192.

Colby, B. G., M. McGinnis, K. Rait, and R. Wahl. 1990. Transferring Water Rights in the Western States: A Comparison of Policies and Procedures. Boulder: University of Colorado, Natural Resources Law Center.

Cordell, H. K., J. C. Bergstrom, G. A. Ashley, and J. Karish. 1990. Economic effects of river recreation on local economies. Water Resources Bulletin 26(1):53-60.

Crandall, K., and B. G. Colby. 1991. Economic Benefits of the Hassyampa River Preserve. Agricultural Economics Discussion Paper. Tucson: University of Arizona.

Daubert, J. T., and R. A. Young. 1979. Economic Benefits from Instream Flow in a Colorado Mountain Stream. Colorado Water Resources Research Institute Completion Report No. 91. Fort Collins: Colorado State University.

Erlenkotter, D., M. Hanemann, R. E. Howitt, and H. Vaux, Jr. 1979. The economics of water development and use. Pp. 169-206 in Earnest A. Engelbert, ed., California Water Planning and Policy: Selected Issues. Davis: University of California, Water Resources Center.

Getches, D., L. MacDonnell, and T. Rice. 1991. Controlling Water Use: The Unfinished Business of Water Quality Protection. Boulder: University of Colorado, Natural Resources Law Center.

Gibbons, D. C. 1986. The Economic Value of Water. Washington, D.C.: Resources for the Future.

Graff, T. J. 1986. Environmental quality, water marketing, and the public trust: Can they coexist? UCLA Journal of Environmental Law and Policy 5(2):137.

Greenley, D. A., R. G. Walsh, and R. A. Young. 1989. Economic Benefits of Improved Water Quality. Boulder: Westview Press.

Higginson-Barnett Consultants. 1984. Water Rights and Their Transfer in the Western United States. Report to the Conservation Foundation, Salt Lake City, Utah.

Howe, C. W., J. K. Lazo, and K. R. Weber. 1990. The economic impacts of agriculture-to-urban water transfers on the area of origin: A case study of the Arkansas River valley in Colorado. American Journal of Agricultural Economics 72(5):1200-1204.

Johnson, N. K., and C. T. DuMars. 1989. A survey of the evolution of western water law in response to changing economic and public interest demands. Natural Resources Journal 29(2):372.

Loomis, J. 1987. Economic Evaluation of Public Trust Resources of Mono Lake. Institute of Ecology Report No. 3. Davis: University of California.

MacDonnell, L. J. 1990a. P. 65 in The Water Transfer Process as a Management Option for Meeting Changing Demands. Report prepared for the U.S. Geological Survey. Vol. I. Boulder: University of Colorado, Natural Resources Law Center.

MacDonnell, L. J. 1990b. Shifting the uses of the waters in the West: An overview. In Moving the West's Water to New Uses: Winners and Losers. Proceedings of the 1990 Annual Summer Program. Boulder: University of Colorado, Natural Resources Law Center.

National Research Council (NRC). 1987. The Mono Basin Ecosystem: Effects of Changing Lake Level. Washington, D.C.: National Academy Press.

National Research Council (NRC). 1989. Irrigation-Induced Water Quality Problems. Washington, D.C.: National Academy Press.

Nunn, S. C., and H. M. Ingram. 1988. Information, the decision forum and third party effects in water transfers. Water Resources Research 24(4):473-480.

Oggins, C., and H. Ingram. 1990. Does Anybody Win? The Community Consequences of Rural-to-Urban Water Transfer: An Arizona Perspective. Issue Paper No. 2. Tucson: University of Arizona, Udall Center.

Saliba, B. C., and D. Bush. 1987. Water Marketing in Theory and Practice. Boulder: Westview Press.

Sly, P. 1988. Reserved Rights Settlement Manual. Washington, D.C.: Island Press.

Stevens, B., and A. Z. Rosen. 1985. Regional input-output methods for tourism impact analysis. In D. Probst, ed., Assessing the Economic Impacts of Recreation and Tourism. Asheville, N.C.: USDA Forest Service, Southeastern Forest Experiment Station.

Tarlock, A. D. 1990. The role of market transfers in the accommodations of new uses: A case study of the Truckee Carson Basin. In Moving the West's Water to New Uses: Winners and Losers. Proceedings of the 1990 Annual Summer Program. Boulder: University of Colorado, Natural Resources Law Center.

Wahl, R. W. 1989. Markets for Federal Water: Subsidies, Property Rights, and the Bureau of Reclamation. Washington, D.C.: Resources for the Future.

Weber, K. R. 1989. What Becomes of Farmers Who Sell Their Irrigation Water? The Case of Water Sales in Crowley County, Colorado. Grant No. 885-054A Report. Ford Foundation.

Wigington, R. 1990. Update on market strategies for the protection of western instream flows and wetlands. In Moving the West's Water to New Uses: Winners and Losers. Proceedings of the 1990 Annual Summer Program. Boulder: University of Colorado, Natural Resources Law Center.

Williams, S. 1990. Winters Doctrine on Administration. Paper on Tribal Administration for Rocky Mountain Mineral Law Institute, Denver, Colo.

Woodard, G. 1988. P. 170 in The Water Transfer Process in Arizona: Impacts and Options. Tucson: University of Arizona, College of Business and Public Administration.

Young, R. A., and S. L. Gray. 1972. Economic Value of Water: Concepts and Empirical Estimates. Springfield, Va. National Technical Information Service, Accession No. PB-210356.

3

The Role of Law in the Transfer Process

What I witnessed for a few hours was the operation of that legal mechanism by which water is prepared for its eventual pumping toward money. It has to be adjudicated, it has to have its claims of ownership documented, it has to have its title quieted, it has to be made merchantable, saleable, which is what enables it to be freed up from land, acequia, community and tradition.

Stanley Crawford, 1990

In the western United States, as in virtually all societies of the world, water is a public resource that is used under rules designed to achieve broad public benefits. Beyond the most basic human needs, it was important in the midnineteenth century to encourage productive use of water in the West, where its availability is seasonally and geographically limited. This was achieved by allowing citizens to use water for private gain and by providing legal protection for those uses. Like a subsidy, the award of water rights to private parties created incentives to encourage the investment and economic activity necessary to meet both regional and national development goals.

Investments in natural resource development sparked both economic activity and settlement in the early West. The land was almost all federally owned, so a system of according legally secure water rights to the first user encouraged investments in irrigation systems to serve homesteads carved out of the public domain and mines on public land. Although the land was federally owned, the government did not dictate how rights to water would be allocated. Instead, it left the settlers to their own devices, and they created a system of appropriative rights to use the water found on the government's land. The earliest user to take water out of the stream and put it to a "beneficial use" acquired the right to continue using it and could prevent others from interfering with the use. This approach was consistent with the government's desire to promote western expansion as well as local desires for economic development. Thus the prior appropriation doctrine was adopted throughout the American West.

As western economies matured, the water rights system proved adaptable to increasing and competing demands. The key to adaptability was that water rights were not restricted to use on a particular parcel of land or a specific type of use. In principle, rights could be transferred from one user to another, and water could be delivered as far as technology and economics could move it.

The ability of an appropriator to transfer a water right—that is, to convey the legal priority to use a quantity of water for a beneficial purpose—is the valuable "property" that the law recognizes in water. A transfer is subject to the condition that a change in use must not injure any other water rights holder. This "no injury" rule is the only universal restriction on water transfers. Some states, however, began to freight the privately held right with other restrictions.

During the early part of the twentieth century, states enacted laws that sometimes frustrated the transferability of water rights. The laws were responses to sentiment that sought to stabilize agriculture and prevent speculation. Some states ruled that water rights were attached to a particular parcel and could not be transferred away from it. Others restricted transfers out of agricultural uses. Together with application of the no injury rule, these restrictions made the transfer process more cumbersome but still worth the effort when the profits from selling were great enough.

A far more serious problem with the transfer process, however, is that several interests historically have been left out of the decision-making processes used to allocate and reallocate water. As discussed in Chapter 2, these include but are not limited to rural communities, ethnic minorities, fish and wildlife and their habitats, and the public. It is now clear that changes in water law and institutions are needed to ensure that all the significantly affected interests, or third parties, are represented in water transfer negotiations.

Indeed, consideration of the interests of parties not directly involved in buying or selling a water right is becoming part of the water transfer process. States are changing their laws in a variety of ways to respond to demands for broader public representation in water transfers. Several federal laws and programs are beginning to address transfer-related problems. The patchwork of state and federal laws acts as a bandage—covering specific issues such as endangered species, water quality, and wetland protection—but fails to provide a comprehensive allocation vision. In addition, some interests have inveigled public agencies or parties directly involved in transfers (sometimes cities or other public entities) to give them a voice in the decisionmaking process and compensation for values lost.

Although the evaluation of third parties in water transfer activities in the West remains incomplete, methods for including them in the process can be gleaned from the experiences in several states. Based on research and the case studies reflected in Chapters 5 through 11 of this report, the committee has identified the following institutional measures and legal authorities for integrating multiple public values and protecting affected interests in the water transfer process:

- public interest review processes;
- impact assessment;
- comprehensive planning;
- judicial public trust doctrine;
- Clean Water Act, Section 404;
- ad hoc negotiation among affected parties; and
- other legislation (including, but not limited to, instream flow laws, area-of-origin protection, water quality laws, conservation programs, endangered species protection, and land use controls).

It seems inevitable that one price of improving water laws will be greater complexity. Yet the need to consider interests important to broad segments of society is so fundamental as to justify some complication of the transfer process. The committee believes that rigorous consideration of public values is necessary and that it is possible without creating major obstructions to desirable transfers.

A goal of modern western water policy, then, is to streamline the systems that impose superfluous restrictions, costs, and delays on the transfer process and, at the same time, to devise new ways to account for important interests that are now left out. These actions may result in a net increase in the transaction costs of transfers, but these costs are justified by the greater public satisfaction and broader public benefits they will ultimately bring. Moreover, a system that accounts for all significant costs will in the end produce a fair and efficient allocation of water. It should encourage transfers with high net gains and discourage those with high third party costs.

As mentioned above, there is no coherent body of law governing how and under what conditions water is transferred. A single transfer can involve several state and federal laws. A transfer in one state can take several years and cost thousands of dollars, whereas an apparently similar transfer in another state can be accomplished quickly and cheaply (MacDonnell, 1990). The various elements of the law of water transfers as it exists now are described below, followed by a discussion of options for improving water law and policy.

STATE WATER ALLOCATION LAWS

Once water is appropriated and a right established, that right generally may be used in other places, it may be used for other than its original purpose, and it may be conveyed to others. Water rights are property rights and they include the right to make changes. As with other property rights, water rights are always subject to redefinition and regulation by state law. From the earliest prior appropriation cases, water rights could be changed only to the extent that no other water rights were affected adversely. Other conditions have been added by some states to achieve other social goals.

There is a general trend in the West of encouraging transferability of water rights to achieve greater efficiency of water use. This trend is visible in the removal of some formal barriers to transfers, implementation of existing laws in ways more conducive to transfers, and, in a few cases, the enactment of legislation to encourage transfers. Legislation that directly involves states in facilitating transfers is rare; California law, however, directs state officials to play an active role in facilitating transfers, and, as noted in Chapter 2, both California and Idaho operate water banks.

Although many states apparently perceive the benefits of water transfers, few have fully considered how best to deal with the negative effects on third parties. Although the issue is not treated with a coordinated approach, several state laws and programs address some of the impacts on affected parties or establish processes to help assess, avoid, or mitigate the effects of transfers. But the overriding legal concern at the state level has been to protect other water rights holders—not third parties—from transfer impacts.

The No Injury Rule

All third party protection schemes build on one fundamental principle, the no injury rule. In the western states, a person seeking to change the use of a water right must request permission from an administrative board, state engineer, or official (or, in Colorado, from a court). The request may come from either the buyer or the seller; sometimes contracts for water transfers are made contingent on getting this approval. Changes in the way water is used, place of use, point of diversion, purpose, or time of use are permitted subject to the condition that the change must not impair uses by other water rights holders. This means that a change in use must not alter the stream conditions that existed when others made their appropriations if it would interfere with others' ability to continue their rea-

The Role of the State Engineer in Water Transfers

In every western state, some public official or entity is responsible for administering the state's programs of allocation of use of water resources. In the early days of settlement, water rights were initiated by diversion and use and by the posting and recording of notices of intent to use water from a stream much in the manner of a mining claim. By the beginning of the twentieth century, this system began to give way to an application and permit system administered by a state official (the "Wyoming Plan"). In many states this person was referred to as the state engineer, although sometimes the official has been designated state water commissioner, state reclamation engineer, chief engineer, or state hydraulic engineer. Today, although the title of state engineer continues to exist only in Colorado, Utah, Wyoming, New Mexico, Nevada, and North Dakota, that office or its equivalent in other states has responsibility for most state water development and use programs.

In general, water rights are property rights and may be transferred as part of the sale of land on which they are used or apart from the land. Typically, applications to obtain approval of a change in point of diversion or place, period, or nature of use of a water right in a western state are filed with the state engineer. Although the procedures may differ from state to state, the purposes are similar. The state engineer examines proposed transfers both to protect existing uses of water from the effects of third party water rights transfers and to protect the public interest.

The state engineer must also ascertain the amount of water that can be transferred, which is generally only the quantity that has been consumed by the prior use. Amounts diverted but not consumed must be left in the stream to protect other users. In making this determination, the state engineer must also consider the impacts of a proposed change on the timing of withdrawal and return of the unconsumed part of the water diverted to a source. Such a change could occur where a right is proposed to be converted from agricultural use to municipal or industrial use. For example, most agricultural water demand is seasonal, whereas municipal and industrial use can be either continuous or seasonal, as is the case with golf courses and other landscape irrigation. Conversion of use without limitations and conditions could result in injury to other users of the resource.

State engineers are quasi-judicial officers with responsibilities to make initial decisions concerning water rights matters. Their actions are governed by state statutes and prior court decisions that have interpreted the law. Statutes are designed to provide an orderly process and to give some measure of security and certainty to established uses of water, thereby protecting the economy of an area. If holders of water rights could transfer the point of diversion or place, period, or nature of use of a right without regard for the effects such a transfer would have on other users, the entire water system of the West would be in jeopardy. To prevent such adverse impacts, the state engineer may impose conditions in transfer applications. These may include, for example, limitations on the duration of pumping, restrictions as to the depth or perforation interval of a well, and limitations as to the time of year in which water may be diverted.

Even in the absence of any injury to other water rights holders, the state engineer in most states has the discretion to deny a transfer application on the basis that to grant it would be inimical to the public welfare of the state or would be contrary to the "public interest." Until recently, public interest review was occasionally used to deny appropriations for inefficient projects or to subordinate a prior right to a larger public project. State engineers are now being asked to use the principle to accommodate the full range of contemporary water uses from irrigation and hydropower to the protection of ecosystems. There are few precedents to guide the state engineer's discretion, and the problem is compounded by the argument that the judicial public trust doctrine requires that the balance be weighted in favor of environmental values.

sonable uses. The no injury rule extends to all appropriators, junior as well as senior, and can be extended to other water use claimants.

Strong as the no injury rule is, objecting third parties are not always fully protected. A change may be allowed if there will be enough water for the objecting water rights holder to enjoy the right. But the rights holder remains vulnerable to call by a more senior rights holder. Before the change, there might have been enough water in the stream to satisfy all rights holders, but now the original objector will have to cease taking water to satisfy the more senior right. This can happen when the change is entirely downstream of the objector if it results in there being less water for a downstream senior rights holder. The senior rights holder need not actually object to the change, because the principle of seniority ensures that junior appropriators are responsible for guaranteeing that senior rights are satisfied. For this reason the objector can assert the increased vulnerability to a call as an "injury" and object to the change under the no injury rule.

Permit proceedings for a change of use provide an opportunity for others to object. Permission may be granted even where injury is incurred, but conditions may be attached to protect other users—such as restricting the amount diverted, replacing water in the stream from other sources, or supplying supplemental water to objecting parties. Sometimes objectors accept cash or other concessions (e.g., headgates, sprinkler or drip irrigation systems, or reservoirs). Appropriators often complain of injury when a proposed change will deprive them of return flows they could use. (Irrigators typically return a large part of the water used for irrigation to the stream via drainage ditches, seepage, and surface runoff, and this water is used by other appropriators downstream.) If a farmer wants to irrigate previously unirrigated lands but not increase the area irrigated, the place of return flows may change. The return flows will no longer be available to another water rights holder who benefited from the return flows at the old location. Thus the change in place of use will not be allowed. One way to limit this type of transfer-related injury is to limit the transferable amount to the past consumptive use only.

Historical Use Limitation

Courts have added a gloss on the no injury rule to take account of the fact that water rights often exist on paper that do not reflect how water actually has been used. Typically in the West, the quantity of water represented by old rights is greatly overstated. Furthermore, rights often are stated in terms of the amount one may divert, even though much of the amount diverted actually has been returned to the stream. Courts and administrative agencies cannot rely on old court decrees for guidance because these are notoriously inaccurate indicators of the amounts of water actually put to beneficial use. Before there were state water agencies staffed with professional engineers to oversee the appropriation process, the quantities appropriated often were based on the "best guess" of the appropriator (which was typically high). Thus in many western states when an appropriator seeks to change the use of a water right, which is necessary to transfer the right away from the original land or to a new use, the amount of historical *consumptive* use must be determined.

The quantity of water that may be transferred to a new use will be limited to the amount of water reasonably consumed. This calculation is based on historical evidence. Because accurate records of the amounts consumed are rare, the state engineer or water agency may consider the type of crops cultivated, soil type, climate, and other factors that indicate water consumption. In some cases, experts are hired; most agencies, however, rely on calibrated equations and climatic data to estimate use for particular locales and particular crops.

The purpose of the historical use limitation on transfers and changes of use is to maintain the conditions that were present when all existing appropriators began their uses. It ensures that they will not be disrupted or harmed by changes. A common problem with transfers from agricultural to municipal use is that irrigation is usually seasonal and municipal demand is relatively constant all year. If a city buys a farmer's water right, the city may be able to take water out of the stream only at the time the farmer historically diverted it. The city may solve the problem by building a reservoir to store water during the irrigation season to be used later.

Appurtenancy Restrictions

Prior appropriation does not require that water rights be appurtenant (legally tied) to the land on which they originally were applied. Several states adopted appurtenancy rules during the reclamation era to promote a stable agricultural society. Under these rules, water rights can be transferred along with the land but not apart from it.

There were several motives for states to adopt appurtenancy rules, which effectively negate the right to transfer or change the use of water rights. It was one way of containing the problem of overstated water rights: the restriction made it unnecessary to deal with the disruption of stream conditions that might result if an artificially large right could be transferred and actually used. This problem is now addressed by the historical use rule, a solution that requires engineering expertise and administrative personnel that were lacking in an earlier era when it was easier simply to prohibit transfers.

Another motive for appurtenancy rules was to protect agriculture. Indeed, most laws restricting transfers or changes in use away from the land relate specifically to irrigation water rights. The laws were an attempt to keep good senior rights in agricultural uses. The rationale was that the rights would be "lost" to farmers if they could be moved to industry. Because agricultural rights were usually so overstated in amount, they were attractive to speculators. Prohibition on transfers away from the land limited the value of the farmer's water right to its utility on the particular piece of land, a substantial modification of the farmer's property right in the water.

Most appurtenancy restrictions have been removed or diluted as the result of economic and social pressures. Most states allow severance from the original land if continued use there becomes economically infeasible. Still, at least five states retain some legal restrictions on transfers away from originally benefited land, although the precise effect of these restrictions is difficult to determine. They are likely only to chill certain potential transfers rather than actually prohibit ones that parties are determined to carry out. These restrictions may have some indirect benefits to rural communities and basins of origin by cutting down the number or kind of transfers, but they do not directly address the third party effects of transfers.

Basin-of-Origin Protection

Moving water out of a region can have obvious effects. The economy, ecology, water quality, lifestyle, and potential for future growth all may be affected. The problem is exacerbated when large amounts of water are moved to another watershed because the basin of origin also loses the benefits of the return flows. A few states have enacted laws designed to restrict the movement of water from one watershed for use in another within the state. Such laws represent a departure from the original appropriation doctrine, which placed no limits on where water can be used. Restrictions to protect the basin of origin apply to new appropriations as well as to transfers of established rights.

California depends on moving large quantities of water from water-rich areas to areas of high demand within the state. The state's area-of-origin protection law gives an exporting area an absolute priority to make future use of water over that of the importing area, and it reserves for the county where water originates all the water it may need for future development. However, as a practical matter, it would be difficult for a county of origin to halt long-standing water exports. The statute provides no procedures and criteria for doing so. Colorado allows conservancy districts to make transbasin diversions out of the Colorado River basin only if they will not impair or increase the cost of present or future water supplies for the exporting area. This has resulted in the building of special "compensatory storage" reservoirs in the Colorado River basin by districts that have imported Colorado River water. More recently, one district has made a variety of concessions and payments to mitigate problems raised by objecting parties in the basin of origin. (See the Windy Gap example discussed in Chapter 6.) Interestingly, there are no similar restrictions on the large cities that import most of the water that is moved between watersheds in Colorado. Montana has a law that requires state participation in all out-of-basin transfers. The size of transfers is limited, and the state is bound to consider public interest factors.

State restrictions designed specifically to inhibit the transfer of water out of state are constitutionally suspect. The U.S. Supreme Court has ruled that because water can be essentially "an article of commerce," states violate the commerce clause if their transfer regulations discriminate against interstate commerce (*Sporhase* v. *Nebraska,* 1982). Thus state regulation of water use must be evenhanded, treating water users within and outside the state basically the same. The state of New Mexico has struggled to fashion a set of constitutionally acceptable restrictions on transfers of ground water that will allay concerns about major appropriations of ground water for the benefit of El Paso, Texas (*City of El Paso* v. *Reynolds,* 1984), and the state has been able to prevent El Paso from obtaining a water right for out-of-state use (Tarlock, 1990).

Public Interest Review

Most states now require that appropriations of water must not be contrary to the public interest. The Utah Supreme Court has held that the same standard applies to a change of use (*Bonham* v. *Morgan,* 1989). Some legislatures have made the requirement explicit. Nevada requires rejection of a transfer application if it would be detrimental to the public interest. To make it more difficult for out-of-state applications to perfect water rights based on the Sporhase case, New Mexico extended to transfers its requirement that appropriations not be detrimental to the "public welfare." In most states, administrative agencies simply apply the same standards relevant to new appropriations, including public interest requirements, to changes of use.

Introduction of consideration of the public interest into the transfer process potentially extends protection to interests well beyond the legal interests of water rights holders. At this point, however, it is not clear how far state agencies will go in recognizing and protecting the interests of people without water rights. The basic idea, a corollary of the principle that all water must be used beneficially, has long been part of western water law, but states have seldom denied new uses or transfers because they contravene the public interest. Sporhase and the new compass of third party interests have stimulated a great deal of interest in the idea, and states are beginning to apply it.

Practice varies from state to state. New Mexico, for instance, has always taken a narrow view of the public interest. When one lower court held that the state engineer should consider the detrimental effects of a transfer on the culture and traditions of Hispanic rural communities, the decision was vacated by the state supreme court and the state engineer returned to his prior practice (In re Application of Sleeper, 1985). Arizona courts have interpreted the public interest as a basis for regulating ground water pumping in urban areas where that is a serious concern (*Arizona Game and Fish Department* v. *Arizona State Land Department,* 1975).

Legislation rarely defines the public interest, much less the process for determining whether it has been adversely affected. In Idaho a vague directive to the director of the Department of Water Resources to determine whether a proposed water use (a new appropriation or a transfer) conflicts with "the local public interest" led the state supreme court to refer to similar language in other Idaho laws and to examine how other states define the public interest (*Shokal* v. *Dunn,* 1985). As a result, administrative hearings are now held to allow affected members of the public to present evidence to the director on matters such as aesthetics, recreation, fishing, and ecosystem functions. The objective is to reach a decision that secures "the greatest possible benefit from the public waters for the public." This involves considering not only the benefit to the applicant but also the economic effect, loss of alternative uses, minimum streamflows, waste, and conservation.

Wyoming is one of the few states with a special review process for transfers. The review looks at economic losses to the community and state that offset benefits from the transfer and the availability of other sources of water. California, through its State Water Resources Control Board, reviews proposed transfers to see if they unreasonably affect the economy of the area from which water is transferred as well as fish, wildlife, and other instream uses.

Instream Flow Protection

In recent years, nearly all western states have enacted laws to protect instream flows. These laws are not targeted at transfers of water rights, but they can ameliorate the negative side effects of transfers by keeping some water flowing in streams regardless of what transfers occur. They may also accomplish the goal of securing instream flows by inhibiting transfers that would deplete a stream below the minimum protected under the instream flow law; instream water rights holders, usually the state, have standing to invoke the no injury rule in proposed transfers.

Some states withdraw or reserve water from appropriation in specific streams. In others, state agencies actually appropriate water to remain in the stream and hold a right to use it for that purpose. Only Arizona and Alaska allow individuals and private organizations to appropriate instream flow rights. (Some individually held rights have been established under an earlier version of Colorado's instream flow law.)

Whether the quantity of water flowing in a stream where water is reserved from future appropriations will be protected from transfers of existing rights depends on the terms and interpretation of particular laws. What is clear is that when instream flow rights are appropriated and made part of the water rights system they will be fully protected against any transfer that would further deplete a stretch of the stream where the right exists. The no injury rule protects instream flow appropriations from the detrimental effects of changes in stream conditions in the same manner as it protects rights for other beneficial uses.

Some instream flow programs are criticized as being ineffectual because the rights appropriated under them are very junior. New appropriations of water for instream flows in streams where existing rights consume most or all of the water do not ensure that there will be water in the stream. But, as explained earlier, even junior rights can be effective in preventing others from transferring or changing the use of senior rights because of the no injury rule. In addition, some instream flow statutes can be used to acquire existing senior rights and convert them to instream flow rights. The Colorado law is being used to accept donations to the state Water Conservation Board of both conditional (unused) and absolute senior consumptive use rights that are converted to instream flow rights. Theoretically, the board could also purchase rights and dedicate them to instream flows.

The effectiveness of instream flow laws in protecting values that are traditionally unrepresented in the transfer process is limited by the scope of the laws. Many are restricted to minimal protection of fish and wildlife habitat and are not used as the basis for securing rights for protection of the ecosystem and other interests such as recreation, water quality, the social integrity of rural communities, or the future economic well-being of a region. In some cases the scope of instream flow statutes is limited by administrative practice, though not by statutory language. Colorado's statute, for example, allows appropriations of only enough water "to preserve the natural environment to a reasonable degree." The board that holds the rights has so far chosen to interpret the law as allowing it to appropriate only enough water for the minimum needs of cold-water fish. Thus it has not sought rights for protection of water quality, riparian vegetation, wetlands, or recreation.

Water Salvage Laws

Because water is so scarce in the West, all intentional and nonintentional uses of water that prohibit reuse merit careful attention. For instance, water lost to antiquated conveyance facilities and inefficient irrigation practices can be conserved. The technology exists to improve these systems, but improvements usually require a substantial investment. The costs of modern sprinkler or drip irrigation systems, concrete ditch lining, laser leveling of fields, and other water-saving improvements can be high in relation to the low profits in much of western agriculture. If the water saved could be put to use by the farmer or sold to others to use, it might justify greater investments in water-saving technology. Under the laws of many western states, however, any water salvaged in this way "belongs to the stream," not to the person who saved it. (See *Salt River Water Users' Association* v. *Kovacovich*, 1966.) That is, the salvaged water can be used by others downstream.

As a simple illustration, consider a farmer with a water right to 1,000 acre-feet (1,230 megaliters (ML)) of water who historically consumed 800 acre-feet (990 ML) and returned the other 200 acre-feet (250 ML) to the stream. By piping the water instead of conveying it in an open ditch, where it can evaporate and seep into a marsh, the farmer can salvage 200 acre-feet (250 ML) of water that was formerly "consumed" by evaporation and unrecoverable seepage. If the salvaged water can be used on a new field by the farmer or sold to a city, it could be worth the cost of investing in the pipe. Without legal assurance of the benefits of the water salvage investment, the farmer has no incentive to change, and less water will be available to others.

California and Oregon have passed laws to encourage water salvage; these generally award title of the salvaged water to the saver. The laws are intended to provide an incentive to water users who reduce the amount of water consumed by undertaking on-site conservation practices: the salvager can sell the water saved to others. Such transfers help to finance conservation improvements and may enable more users to benefit from the same overall quantity of water.

The most celebrated example of water salvage resulting in transferable water is the Metropolitan Water District-Imperial Irrigation District (MWD-IID) transaction in California. (See the Imperial Valley case study, Chapter 11.) The MWD agreed to invest more than

Examples of Water-Saving Legislation

Given the history of western water law, incentives will be necessary to induce water conservation. To date, most transfer incentives are negative and involve fear of losing the right by judicial or administrative action. Several states have recently created positive incentives that seek both to encourage water conservation and to recapture some of the same water for public uses. For example, Oregon has enacted an innovative water conservation plan that splits the benefits of conservation between the saver and the state. If the state approves a conservation plan as feasible, effective, protective of junior rights, and consistent with the public interest, the saver is entitled to the saved water, but the state may allocate 25 percent of the water to itself for instream flow maintenance and other environmental uses (Oregon Revised Statutes 537.470). The legislation raises taking issues, but the argument has been considered and rejected (Sax, 1990).

California has had a water salvage law since the early 1980s. Washington State enacted a similar statute in 1989 and broadened it in 1990. The Saved Water Act (Revised Code Washington 90:38.005-902) applies only to the Yakima basin. Net water savings may be voluntarily transferred to the state's Department of Ecology in trust for instream or irrigation uses. The purpose of the act is to reallocate water for fishery maintenance in the Yakima River, but the act has not been implemented.

$100 million in lining irrigation canals and other projects to salvage water within IID. In return, MWD received the right to use the water saved, some 106,000 acre-feet (130,000 ML) per year, for 35 years. This case illustrates the kind of water salvage transactions that may be possible throughout the West, although its large scale is probably unique. It is also important because, while California had a salvage statute, no special statutory authority was necessary for the transaction to take place.

The purpose of water salvage laws is surely salutary, but fuller use of water from a stream can have negative environmental consequences, especially when large-scale water salvage becomes a major supply augmentation strategy. Streamflows may be depleted, wetlands may dry up, and water quality may degrade as return flows and seepage are reduced. In recognition of this fact, Oregon included in its law a requirement that a salvager must dedicate 25 percent of the water salvaged back to the state to remain in the stream. The MWD-IID trade is an example of a situation in which confronting the environmental consequences has been deferred. Although the Salton Sea is already increasing in salinity, the MWD-IID transfer could accelerate that trend, harming fish and wildlife. Economies supported by the sea may be altered. The environmental issues, however, have not yet been fully addressed because the transfer does not immediately change the fate of the sea.

OTHER STATE LAWS

Besides provisions in water allocation laws that relate to transfers, states have a number of other laws that can affect whether and under what conditions water transfers may be made.

Water Quality

Although all states have water quality protection laws and programs, few of these can be used to restrict transfers in any way. The law of water rights is almost exclusively concerned with quantity, not quality. It is difficult to integrate quantity and quality considerations because of the way in which these separate water use regimes operate. The failure to integrate these two doctrines results in water quality effects that can no longer be ignored.

Generally, water quality is protected by federal law administered by the states. Point source discharges require permits that limit the amounts of particular pollutants by incorporating federally set standards and other limitations as necessary to protect the quality of the receiving stream. This regulatory scheme, targeting point sources of pollution through a permit process, essentially ignores degradation of water quality that results from other causes such as depletion and nonpoint sources of pollution.

Plainly, water quality can be affected dramatically by pollutants from other than point source discharges. The Clean Water Act defines point sources as not including irrigation return flows, yet these are capable of loading considerable sediment, agricultural chemicals, and naturally occurring trace elements into the West's waters. Following the spirit of the law, the courts generally have found that releases from reservoirs are not point source discharges, even though they may greatly alter the temperature and chemical content of water. Throughout the West, water is degraded in quality simply by being used, whether or not pollutants are added. Removing water from a stream reduces its dilutive capacity; storing it in a reservoir changes its temperature and chemical content; return flows from

Water Quality Effects of Water Use

The law of water rights deals almost exclusively with quantity, not quality. Yet in nature, quality and quantity cannot be so easily separated. Diversions and ground water pumping can affect the quality of streams and aquifers by increasing the concentration of contaminants. For example, diversion for irrigation from the Colorado River and its tributaries affects not only the amount remaining downstream but especially its salt content. An interstate compact allocated rights to use water, but later, when quality issues became severe, separate legislation was necessary to address salinity. A treaty with Mexico protected a share of water but not its quality, so that another agreement was needed to deal with the salinity of the Colorado River. In the San Joaquin Valley of California, water quality issues are by far the most challenging of the water issues even if the current drought also stresses the resource in terms of quantity. Transfers of water could alleviate some of the drainage disposal problems; they also, in some circumstances, could adversely affect the ability of streams to absorb drainage discharges.

It can be difficult to integrate quantity and quality considerations because of the ways in which these aspects of water use are regulated. Prior appropriation, in theory, allocates the yearly flow of a stream until it is dry. Water quality is protected through the Clean Water Act, a federal law administered by the states. Point source discharges must obtain discharge permits that conform to federal discharge and water quality standards. Nonpoint source discharges, which include irrigation return flows, are subject to state programs that seek to induce the adoption of best management practices, thus far largely ineffectual.

The problem is further complicated because effluent can be a resource as well as a contaminant. The technology-forcing focus of the federal Clean Water Act seeks to reduce all effluent discharges, and this reduces available supplies of water that are in fact suitable for irrigation in many cases. In contrast, diversions and water use apart from any discharges may increase contamination as pollutant concentrations increase.

The lack of integration of water quality and water allocation goals thus sometimes allows transfers that ignore water quality effects. Existing procedures do not provide a way to protect third party interests in water quality except if the transfer results in harm to third party water rights holders.

When water quality enters into the equation for water allocation decisions, it is generally through public interest considerations. For historical reasons, there are rights in water quantities or flow rates, with quality considerations secondarily applied through regulation of discharges. In a physical or economic context, it would be more rational if quantity and quality attributes were equal partners in the determination of a right, as they must be in the assessment of value (Getches et al., 1991).

any use are changed in quality. There are almost no laws to restrict the quantity or manner of use of water, even when it causes a deterioration in quality. In addition, where water quality laws are in potential conflict with an unfettered right to appropriate water, state laws often mandate that the right to use water shall take precedence over protection of quality (Getches et al., 1991).

The potential water quality effects of transfers are considered in a few states as a part of the public interest review process, but this process may be ineffective because the states typically separate their administration of water allocation and water quality. The permitting process for a transfer or change of use remains a largely unexercised opportunity for pursuing those goals, one that could help to ensure that other users of the stream and the public generally are not harmed by the proposed transfer.

Land Use

The effects of transporting water or transferring a water right could be considered under the existing authority of most state land use planning laws or in the context of local zoning authority. Rarely do state or local governments use the occasion to delve into the consequences of transfers that furnish water for a new subdivision, that require a water diversion facility or pipelines, or that result in the retirement of irrigated farmland. These effects are not considered because normally "water planning" is off limits to land use planners.

Arizona, however, restricts new development in areas of severe ground water overdraft unless the developer can demonstrate a 100-year "assured water supply." In Colorado the consequences of water development can be considered under local government land use authority. Colorado leaves virtually all land use decisions to counties. The counties have specific authority to regulate several development activities if they determine them to be "areas of state concern." One of these potential areas is the construction of water pipelines and facilities. Recognizing that major transbasin diversions can have significant effects on their environment, lifestyle, and economic future, some counties in western Colorado have declared water facility construction to be an area of state concern subject to county permitting. Using this authority, they have held extensive hearings to assess the effects of proposed water projects. For example, Eagle County denied a permit sought by Colorado Springs and Aurora to construct facilities that would dewater a large wilderness area in the tourism-dependent county. (See Chapter 6.)

Environmental Impact

California and Washington have laws requiring the assessment of impacts of major developments, and these can be applied to water rights. For instance, the California Environmental Quality Act requires the preparation of an environmental impact report (EIR) for any proposed major project. The EIR is to enable decisionmakers to balance project benefits against environmental and other costs. The statutory requirement applies to any major water use application, including a transfer.

Soil Conservation

One of the problems of drying up agricultural land is the loss of topsoil through erosion by natural forces. Several counties in the West have enacted ordinances prohibiting blowing dust. These laws operate against landowners. Where land is dried up to transfer the water to municipal uses, the city may be responsible for the blowing dust offense if it purchases both land and water, and the farmer may be responsible if the city purchases only the water. For instance, Arizona state law attempts to cope with this problem by requiring farmland retired from irrigation to be maintained free of weeds and dust.

Properly drafted and enforced state or local laws designed to protect soils could encourage transactions with less destructive effects. Although these laws would not regulate transfers directly, they could be a disincentive to sales that leave farmland abandoned. Faced with required stewardship over the land, the parties may decide on temporary transfers, limited to dry years, or transfers of part rather than all of the water rights, or they may seek other ways either to keep the land in cultivation or to rehabilitate it for other uses.

FEDERAL RECLAMATION LAW

As noted in Chapter 1, the federal government has the potential to shape the future of western water allocation because it stores and delivers large quantities of the region's water. However, as many of the guests who spoke to the committee observed, the federal government's role at the present time is largely passive. This may seem ironic in light of the great influence the Bureau of Reclamation, together with the U.S. Army Corps of Engineers, has exerted in shaping the landscape and economy of the modern West. Through most of the history of development in the West, there has been great tension between the application of state laws and the application of federal laws to the allocation of the West's water. On the one hand, the supremacy of federal over state law is written into the U.S. Constitution. On the other, allocation of water in the western states, as elsewhere in the Union, is generally a state prerogative because of a long history of the minimal assertion of federal powers and interests. Despite its great constitutional power to control resources, the federal government has chosen to defer to state water allocation decisions unless a clear federal policy was at stake. Today, neither the Bureau of Reclamation nor the Corps of Engineers is trying to advance a clear congressional or administrative mandate other than to fulfill statutory project objectives and honor beneficiary obligations. Rather, these agencies are just beginning the long process of redefining their missions in light of modern needs and values.

When Congress created the Bureau of Reclamation in 1902 and authorized construction of large federal water projects throughout the West, it attempted to resolve the conflict between state prerogative and federal supremacy by declaring in Section 8 of the Reclamation Act that

> [N]othing in this Act shall be construed as affecting or intending to affect or to in any way interfere with the laws of any States or Territory relating to the control, appropriation, use or distribution of water used in irrigation, or any vested rights acquired thereunder and the Secretary of the Interior, in carrying out the provisions of this Act, shall proceed in conformity with such laws. . . .

For nearly half a century, as the Bureau of Reclamation built projects in the various states, Section 8 and the federal-state law dichotomy generated little controversy. The states were glad to have the federal projects and through political means had substantial input into the distribution of their benefits and costs. By the mid-1950s, however, significant controversies arose, and, in a series of decisions involving the Friant Unit of the Central Valley Project, the U.S. Supreme Court extended the reach of federal law and federal bureaucratic discretion in the face of challenges grounded in allegedly contradictory state law. This legal trend, however, came to an abrupt halt in 1978 in *United States* v. *California*. Without quite overruling its earlier cases, the U.S. Supreme Court upheld California's authority to regulate the operation of a federal project to ensure environmental protection objectives. In that case, involving the operation of the New Melones Dam and Reservoir, state regulation was approved, subject only to the caveat that it not be inconsistent with a "specific congressional directive."

Arguably, therefore, if a water user with a contractual right to receive water from the Bureau of Reclamation seeks to transfer that right to another, the transfer would be subject basically to state law (Roos-Collins, 1987). Thus it would be allowed, given approval under the laws of the state, unless it were found to be inconsistent with a specific congressional directive. In practice, however, determining what are the relevant specific congressional directives may be a daunting task not only for federal administrators in the U.S. Department of the Interior (DOI) and the Bureau of Reclamation, but also for potential reviewing courts when controversies erupt. The resulting legal uncertainty has in turn meant that, despite the courts' apparent tilt toward the primacy of state law prerogatives in managing federal projects, federal project managers still have broad discretion in determining whether to approve transfers of water under their physical control (Willey and Graff, 1988).

After lengthy internal review, on December 16, 1988, DOI itself sought to place bounds on the exercise of this discretion by issuing a document entitled "Principles Governing Voluntary Water Transactions That Involve or Affect Facilities Owned or Operated by the Department of the Interior." Responding to an expectation that DOI would increasingly be asked to approve or facilitate voluntary water transactions and to the support for water transfers expressed by the Western Governors' Association (July 7, 1987, and July 12, 1988, resolutions in Western Governors' Association Water Efficiency Task Force (1986) and Western Governors' Association Water Efficiency Working Group (1987), respectively), the principles for the first time enunciated a formal departmental policy in support of voluntary water transfers. For example, "DOI's role will be to facilitate transactions that are in accordance with applicable state and federal law and are proposed by others," and "DOI will refrain from burdening [such] transaction(s) with additional fees, costs or charges. . . . " On the other hand, DOI's policy recognized, as have many others, that support for voluntary water transfers cannot and should not be unqualified. The seven principles are replete with qualifications addressing, among other concerns, third party effects, federal contractual and financial interests, and environmental impacts. The DOI policy also generally is deferential to the primacy of state law and water management. It takes the position that in most circumstances DOI should pursue a reactive, rather than proactive, posture regarding specific potential voluntary water transactions.

A document providing further "criteria and guidance" explicating the December 16, 1988, policy was issued for the use of the regional directors by the Commissioner of Reclamation on March 16,

1989. Again recognizing the difficulties caused by overgeneralizations in setting western water transfer policy, the document provides additional instructions for DOI personnel as to how they should approach potential transactions. Notably, it acknowledges that some transfers may require new federal authorizing legislation, that others face protracted and costly legal procedures, and that there may be problems under the Reclamation Reform Act of 1982. Remarkably it does not address a key emerging problem: What should DOI policy be when contracts come up for renewal and all, or more likely some, of the water under contract has been transferred? The document does confirm, however, that DOI will not create a disincentive to transfers by charging the participants to a transaction for "any profit that might be envisioned as the difference between appropriate costs and the market value of the water."

Taken as a whole, the principles, criteria, and guidance place DOI in a posture of measured support for voluntary water marketing. Marketing enthusiasts can take solace from the fact that a formal supportive policy finally emerged from DOI after much internal struggle and reported external pressure. Marketing skeptics, on the other hand, will find relatively little in the policy to criticize. Some may oppose granting federally subsidized water users the right to sell that water at a profit. Others may question DOI's commitment to mediate the adverse environmental impacts of marketing transactions. Still others may prefer that DOI not encourage water marketing at all. For the most part, however, the DOI policy retains sufficient ambiguity and DOI personnel retain sufficient discretion that substantial variability in future departmental treatment of proposed transactions is likely (Wahl, 1989). More transactions probably will be facilitated than would have been without the principles, criteria, and guidance, but the DOI policy by no means provides clear sailing for western water transfers involving federal facilities or federal water rights.

Congress has not yet spoken in generic terms on the law of federal water transfers. Although transfers have been an integral part of recent legislation settling the water rights of various Indian tribes—and, in 1990, protection of the interest of a federal wildlife refuge (see Chapter 5)—Congress has not yet put its imprimatur on any general statement of law either approving or limiting federal water transfers. Given the federal financial interest in the projects it has funded and the national interest in the environment and in the well-being of Indian tribes toward which the nation has a trust obligation, this inaction is difficult to justify. No doubt, however, a full-scale congressional debate on these issues would expose a variety of points of view and pose significant problems for the sponsors of comprehensive water transfer legislation.

FEDERAL ENVIRONMENTAL LAWS

Much of the federal interest relevant to water transfers is asserted indirectly through federal environmental laws. The National Environmental Policy Act requires that an environmental impact statement be prepared anytime a federal action significantly affects the environment. This requirement is most likely to involve the Bureau of Reclamation in the operation of one of its projects, but it may also involve the Bureau of Land Management, the National Park Service, or other federal entities when federal lands are involved.

Where wetlands are implicated, Section 404 of the Clean Water Act may come into play. This provision requires that a comprehensive public interest review be conducted by the U.S. Army Corps of Engineers. The U.S. Environmental Protection Agency also retains the power to veto a Corps of Engineers approval of a project affecting wetlands, a power that it exercised to halt the Two Forks Dam project in Colorado and that it conceivably could use in a water transfer situation not only to prevent or mitigate a loss of wetlands but also to fulfill various objectives under an expansive notion of the public interest.

The Endangered Species Act is another federal statute that may affect proposed water transfers in the West. As understanding of the broad range of environmental impacts caused by existing projects on the streams of the West has increased, concern for the protection of native species and their habitat has mounted. Federal agencies must consult with the U.S. Fish and Wildlife Service to determine the effects of a project on endangered species habitat and are flatly prohibited from proceeding with any project that would jeopardize the survival of a listed species or its habitat. This prohibition may restrict the ability of water users to divert water from streams. Even in large basins such as the Snake-Columbia, the Colorado, the Sacramento, and the San Joaquin, endangered species protection may become a binding constraint not only on new development proposals but also on proposed water transfers that otherwise would make water management in the basins more efficient.

The challenge for federal water managers, as well as for federal wildlife managers and environmental regulators, will be to find methods for improving environmental conditions even as water is being transferred among consumptive users. The case studies in Chapters 5 and 8 examine several situations in which this has been done or is being attempted. The first step is to accord formerly excluded values such as environmental quality equal or increased weight relative to traditional uses of water and to create processes that involve a wide range of interested parties in making the trade-offs among these values.

One basin-specific model for resolving resource conflicts is the Northwest Power Planning and Conservation Act of 1980. The act was passed to aid in the development of hydroelectric power in the Columbia River basin and "to protect, mitigate and enhance the fish and wildlife habitat of the Columbia River and its tributaries" (16 U.S.C. §839). The Columbia River basin covers an area larger than France, draining parts of seven states and British Columbia. The act establishes a council broadly representative of federal, state, and tribal interests in the region that oversees related basin development and conservation activities. The council was given responsibility to develop a power plan and a fish and wildlife program to protect, mitigate, and enhance fish and wildlife affected by the development and operation of hydroelectric facilities while ensuring efficient and reliable power supplies. The council, which is charged to encourage broad public involvement, serves as a regional forum, outside the courts, for disparate interests to reach agreement on resource use decisions. Over the past 10 years, the council has helped establish facilities to help migrating fish pass dams on the mainstream of the Columbia and Snake rivers, encouraged conservation programs for utilities that have resulted in substantial cost savings, and generally provided regional leadership on power planning. Although this mode of activity faces significant future challenges—especially related to salmon runs—it appears overall to be an effective mechanism for improving resource decisionmaking.

FEDERAL AND INDIAN RESERVED RIGHTS

Until the 1970s, much of the tension between federal and state water allocation centered on the proprietary claims of the federal government—which in effect could trump state-created rights. Federal water rights can be claimed for Indian tribes and certain reserved public lands. Of the two, Indian water rights are potentially the most disruptive.

The Winters Doctrine

When Indian reservation lands were set aside by treaty, statute, or executive order, the reservation included sufficient water to fulfill the purposes of reserving the lands. The U.S. Supreme Court ruled in the 1908 case of *United States* v. *Winters* that the reservation of land implies a reservation of water even though the document establishing it is silent as to water; otherwise the purpose of providing a permanent homeland for the tribe could not be accomplished.

Application to Federal Lands

Much later, the U.S. Supreme Court held that the same reservation of water rights that attends reservations of Indian lands attends the reservation of federal lands for purposes such as national parks, forests, wildlife refuges, and military bases. The quantity reserved in each case is determined by looking at the basic legislation creating the land system and the legislation setting aside the particular tract. Because of the nature of federal resource protection systems (wild and scenic rivers, parks, and so on), most reserved rights are likely to be used to maintain instream flows rather than for consumptive uses. Of course, military bases, oil shale reserves, and visitor centers or campgrounds in parks can claim water for consumption.

Priority of Reserved Rights

Where federal (or Indian) rights exist, they are incorporated in the hierarchy of rights in the state system according to priority. They are not, however, lost when they are not used. The priority date is the date when the reservation was set aside, not when water was first used on the reserved land. Reserved rights must be respected whenever transfers are made by private parties; reserved rights are protected from harmful transfers by the no injury rule just like other rights in the appropriation system.

Quantification of Reserved Rights

Federal law makes the existence and quantity of reserved rights subject to adjudication in state courts. The quantity is determined by the court's determination of how much water is needed to fulfill the reservation's purpose. These purposes are usually found by examining the words, circumstances, and intent on both sides of the treaties and agreements setting up the reservation. The quantity to which a tribe is entitled can also be determined through negotiations. Several tribes are negotiating with states and local water users to set the amounts of water to which they are entitled. Settlements in Colorado, California, Montana, Nevada, Idaho, Arizona, and elsewhere have been cemented by agreements approved by the courts or Congress. Whether litigated or negotiated, the resulting quantity of the right becomes part of the overall hierarchy of rights of all water users on the particular stream.

As defined by the state's appropriation system, the U.S. Supreme Court found that where a purpose of an Indian reservation is to encourage Indian agriculture, rights to accomplish that purpose are to be determined on the basis of the amount of practicably irrigable acreage within the reservation. This formula was applied by the Wyoming Supreme Court in the Big Horn adjudication of the rights of the Wind River Reservation tribes. Once they are quantified, however, Indian reserved rights may be put to any beneficial use. Thus rights quantified on the basis of the amount of practicably irrigable acreage within a reservation can be applied to mining, municipal uses, irrigating a resort's golf course, or even maintaining instream flows for a fishery. Other methods of quantification may be used where the purposes of the reservation were different (e.g., to maintain a fishery or to establish "a permanent home and abiding place"). Because agriculture and most other purposes for which reservations were established demand large quantities of water, Indian tribes can claim rights to a significant share of the water in many western rivers. The future viability of reservations as tribal homelands depends on what use is made of this water. For many tribes, water is essential to economic and cultural survival. Throughout the West, the entitlement to water under the *Winters* doctrine is being translated into quantified rights and into actual supplies available for use on reservations. In the meantime, non-Indian water users are using water supplies allocated to them under state water laws that may eventually be judged to be tribal water.

Transferability of Indian Reserved Rights

Agriculture and fishing depend on adequate supplies of water. But some tribes are not traditionally farmers or fishers and are located where such enterprises are futile or at least uneconomical. Others need some water but could gain from leasing a portion of the water to which they have rights to non-Indians. Thus several Indian tribes believe that the most productive and profitable uses of their reserved water rights are off the reservation. Marketing water is one way to use their reservation resources to become self-sustaining. Indians thus may seek to market their water rather than use it for capital-intensive, low-profit agriculture. This practice is also encouraged by non-Indians in some cases because they have become dependent on using undeveloped Indian water. If the tribes begin using water on their reservations, it could disrupt established non-Indian economies or at least require them to seek new water sources.

Despite the possible mutual benefits of many transfers of Indian water to off-reservation uses, objections are raised from time to time on both sides. Some Indians fear that conveyance of the right to use their water will limit future use of reservation lands and constrain their opportunities for cultural survival or economic growth. The fear is not without substance and argues for careful, well-informed tribal decisions.

Non-Indians sometimes object that transfers of Indian rights could result in tribes' charging for the continuation of a supply that non-Indians are now getting for free. As long as the Indians lack capital to develop the water that is nominally theirs under reserved rights—funds for irrigation systems, storage reservoirs, and the like—the water is available to non-Indians. Some objections to proposals for compensated transfers are based on the argument that Indian rights can be used only on the reservation and any transaction that provided for off-reservation use would be unlawful as inconsistent with the purpose of the reservation. Of course, anytime Congress approves a transaction, it should moot the question of whether off-reservation use is permissible.

The legal power of tribes to transfer water off the reservation remains unresolved in the courts, and Congress has chosen to deal with the issue in an ad hoc manner. Congress has approved a number of transfers of the right to use Indian water off the reservation in recent years in the context of negotiated settlements of Indian reserved water rights claims with states and non-Indian water users. Nearly all these settlements included provisions for limited marketing of Indian water. Early in the nation's history, Congress passed the Nonintercourse Act, making property transactions with Indians unlawful unless the United States approves or is a party. The act was intended to protect the tribes' perpetual rights in their lands. It also preserves the federal government's prerogative, exclusive of other nations and of the states or private parties, to extinguish Indian title to property when transactions occur and to regulate trading with Indians, a prerogative expressed in the commerce clause of the Constitution. It provides an opportunity for Congress to exercise what has been described in a number of Supreme Court decisions as a fiduciary relationship toward the Indians. Thus the United States oversees these transactions as a trustee for the tribes. The federal role in Indian property transactions and trade does not imply cutting off the tribes from commerce with neighboring economies. Indeed, reservations may depend on access to non-Indian markets for farm goods, minerals, and industrial products. But the fact that Indian property is held in trust by the government does affect the legal requirements for transferring Indian lands and interests in land, including water. The full reach of the Nonintercourse Act has not been determined.

At this point, transfers of rights to use Indian water by lease, sale of a quantity of water, or agreement of a tribe to defer using water so that it is available to others may not be secure without congressional approval. Many experts have urged legislation generally authorizing the lease of Indian reserved water rights to non-Indians. Congress has not, however, enacted such a statute. One possible enactment would be to allow tribes to lease water rights in a manner comparable to the leasing of reservation lands, which may be done under a statute requiring the approval of the Secretary of the Interior. If Congress passes such a law, it should include a process for reviewing the potential effects of such transactions on the reservation culture, economy, and ecology. It should be noted that (1) if the courts were to hold that Indian reserved rights cannot be transferred off reservation under existing reserved rights and (2) if a given transfer would adversely affect other holders of water rights, then (3) it presents a serious constitutional question whether Congress could grant the right to make such transfers in the absence of compensation to the losers.

SPECIAL DISTRICT LAWS AND STATE LAW RESTRAINTS

There are literally dozens of types of entities that develop and distribute water to individual water users. The earliest North American examples are the acequias, the community ditch organizations of the Southwest that are rooted in ancient Spanish custom. Many still operate in northern New Mexico, where they not only perform water distribution functions but also are often at the center of community life. Private entities that distribute water include mutual water companies and irrigation companies. In a mutual water company, rights are owned by the individuals who make up the company. Other companies hold rights directly, with the ownership interests of rights to use water evidenced by shares of stock. Although every type of water rights holding entity has a potential role in water transfers, the largest and most influential types of public entities that exercise water rights in the West are irrigation and conservancy districts.

About a thousand entities known by a variety of names—irrigation districts, conservancy districts, water authorities, and the like—supply a large measure of the water in the West, most of it to agriculture. About one-third of all irrigated land is served by these districts, but a much larger percentage of western water is under their control. In many cases, they manage federally financed projects and distribute the water from those projects to district members. The water rights held by a district belong to the district itself; members or residents of the district have contractual rights to use water.

Districts have considerable promise for participating in and improving water transfer activities. Some districts have programs that move water freely within their boundaries according to the annual and seasonal needs of farmers and others, including municipal users. A well-known example is the Northern Colorado Water Conservancy District, which maintains a robust market, annually trading contractual rights to use district water among district members. (See Chapter 6.)

The ability of special districts to transfer water often is constrained by state law. Many districts cannot transfer water beyond their own boundaries. In the Colorado Front Range, for example, the Northern Colorado Water Conservancy District holds far more water than it can use in the foreseeable future. As taxpayers, district residents must repay the high costs of developing the water, but the district and its member entities (irrigation companies and municipalities) are prohibited from selling it to growing municipalities surrounding the district who need water now. These water-short communities must then seek water from costly and sometimes controversial distant sources, such as transbasin diversions or expensive and environmentally questionable facilities like Two Forks Dam. Meanwhile, the district has sought ways to attract municipal growth to provide an in-district market for water. But seeking growth brings the district into conflict with some established municipalities within it, which have land use policies favoring slow growth to maintain the quality of life.

Some states are addressing the tensions created by restrictions on out-of-district transfers by lessening or eliminating the restrictions. Colorado has authorized conservancy districts to lease water for use outside their boundaries, and Utah has removed all restrictions on transfers out of districts. With the lifting of transfer restrictions, water may move more freely to other areas, but there is no assurance that interests other than those of water consumers will be considered. Although absolute restrictions on district transfers err on the side of inhibiting beneficial transactions, there are sound reasons to examine carefully the consequences of removing water from one area and exporting it to another when public as well as private interests are at stake.

Given the large amount of water controlled by special districts, it is important that state law address the question of what entity or process should decide when, how much, and on what conditions water should move from one region to another. In Arizona, all transfers out of districts or watersheds where districts develop their water, even water not subject to district water rights, are subject to the veto of affected irrigation districts, agricultural improvement districts, and water users' associations. Yet these entities are charged with narrow purposes and were never intended to represent the social, economic, and ecological interests of a region. Often their governing boards are neither elected nor appointed in a politically responsive process. Decisionmaking on these matters would seem more appropriate at the county, regional, or state level. Nevada requires that review and comments be sought from county commissioners when a transfer that will remove water from the county is proposed, but recommendations are not binding on state administrative officials. The state public interest review statutes that exist in most western states are a potential, but largely unused, means for considering public issues surrounding water transfers from special districts.

Districts are charged with responsibilities broader than the traditional functions of water development, conveyance, and storage, such as playing a role in conserving water and in preserving water quality. In some states, districts are assuming responsibilities for protecting public interests besides water supply. In Colorado, conservancy districts can administer programs to control nonpoint source pollution. Unless districts are given broader responsibilities than simply ensuring water supply, they cannot be expected to exercise their potential powers over transfers to accomplish other goals.

OPTIONS FOR IMPROVING WATER LAW AND POLICY

Public Interest Review Processes

Even though all but one western state use some type of public interest review process in water decisions, all could improve their manner of reviewing the effects of water transfers. This is particularly true in state processes that concentrate the public interest review on new appropriations and give little attention to transfers. Most states lack clear standards for defining the public interest to be protected. Many of the social, economic, and ecological interests that are affected by moving water from one use or region to another are not included. Once the elements that constitute the public interest are identified, decisionmakers need state policy guidance to resolve conflicts among competing interests. Comprehensive water planning can serve to articulate both the elements of public interest and the state policies relating to them.

Public participation in review processes is essential in identifying the range of interests affected by a proposed transfer. Procedures vary widely from state to state, but most western states provide opportunities for people representing all the elements of the public interest as defined by state law to be heard. A few states, including Montana and Utah, still limit participation in transfer proceedings to those who hold water rights.

Impact Assessment

The National Environmental Policy Act requires the assessment of potential environmental impacts when major projects are undertaken. It applies to any proposal requiring federal approval or license or the use of federal facilities. A few western states, including California and Washington, have adopted laws establishing a parallel set of environmental review requirements for state-permitted or state-sponsored projects. State or federal laws requiring assessment of impacts are important devices for evaluating the effects of a proposed water transfer. The information is valuable in a fair and comprehensive public interest review. States lacking impact assessment programs could fill a significant gap in water decisionmaking by enacting such laws.

Comprehensive Planning

Most western states have some type of water plan. Few, however, have comprehensive planning processes that articulate established water policies in the context of issues that are likely to arise in transfer proposals. The policies and standards developed through comprehensive policy planning can inform the process for public interest review as well as processes for issuance of water rights permits and the development of projects. Planning processes that are essentially dynamic policy development programs exist in Kansas, Montana, and Oregon.

Plans can cover the panoply of values and interests that are affected by water development, transfer, and use. They can discuss the relative importance to the state of water-related values and their impacts on rural communities, fisheries, wetlands, recreation, drinking water, and flood control. They can be designed as guidance documents for water rights holders and decisionmakers in transfers as well as other major water decisions, such as new appropriations, development projects, water quality standard setting, and instream flow protection programs. Thus predictability can be enhanced, costs minimized, and the full range of effects of transfers anticipated in a comprehensive state planning process.

Facilitating Third Party Participation in the Water Transfer Process

State and tribal governments that make the rules regarding who may participate in the water transfer review process can go to varying degrees of effort to facilitate third party participation. An essential follow-up to decisions regarding which interests should be allowed to influence the process involves determining the appropriate degree of influence and considering how their participation can be made more effective. Government efforts to facilitate third party participation can include the following elements:

1. *Permission to be present and speak.* The opportunity for public interest groups to be present at hearings and to state their opinions is implied in state laws that list various factors a state official or agency may consider in reviewing a proposed change in water rights—recreation, local economics, fish and wildlife, and so on.

2. *Legal ability to influence transfer conditions and to delay transfer approval.* Without some bargaining power a third party interest cannot effectively influence the outcome of a transfer review process. Third parties' objections "count" when they can delay transfer approval, influence transfer conditions, or make the process more costly for transfer proponents. Transfer proponents then have an incentive to negotiate with third parties and seek to accommodate their interests. Other water rights holders clearly have such bargaining power under the no injury rule common to the western states. Third parties who do not hold water rights depend on general environmental and other regulatory statues to give them standing to object to transfers.

3. *Designated representation.* A government agency can be assigned the task of representing an interest. For instance, some state game and fish departments have been assigned some responsibilities for assessing transfer impacts on instream flows. However, if the interest may be represented only by a specific government unit, this can effectively limit participation on behalf of that interest. Environmental groups may be especially effective representatives in a discussion of the environmental impacts of a proposed transfer.

4. *Financial and legal assistance.* Economically disadvantaged third parties may need financial assistance to conduct investigations, to collect evidence regarding transfer impacts, and to hire attorneys or other experts. For instance, the federal government, in the context of its trust responsibilities to tribes, provides financial and legal assistance to tribes involved in litigation and

negotiation over water issues that affect tribal interests. Some state water agencies take an active role in preparing assessments regarding transfer impacts on other water rights, and this role could be extended to other third party impacts.

Broader third party participation, at any of the levels described, will increase state and tribal government costs of reviewing proposed transfers, as well as transaction costs incurred by transfer applicants. States and tribes need to arrive at a balance between these increased costs and the benefits of broader and more effective third party representation. The federal government will also have to strike such a balance when specific policies and criteria are drafted for reviewing proposed transfers of federal project water.

Judicial Public Trust Doctrine

The public trust doctrine allows a court to reallocate vested water rights to protect trust values, which now include environmental protection. As applied, it prevents the allocation or transfer of water without adequate consideration of the consequences. It recognizes that water is, at base, a public resource and that private interests in it should further, not impair, public benefits from the use of water.

The doctrine has its origins in civil and common law doctrines that recognize public servitudes (rights of passage) in tidal navigable waters and state ownership of the beds underlying navigable waters. In the nineteenth century the doctrine was expanded to place outside limits on the widespread practice of state grants of submerged lands so that they could be filled in. California has a long tradition of aggressive use of the trust to subject submerged lands granted by the state to private individuals to public rights of passage. These rights include commercial navigation, recreational use, and environmental protection. In 1983 the California Supreme Court applied the public trust to a lawsuit to curtail Los Angeles' appropriations from streams that fed Mono Lake (National Research Council, 1987) to preserve the lake's fragile ecosystem. *National Audubon Society* v. *Superior Court* (1983) held that the vested water rights of the city of Los Angeles could be retroactively limited to support trust values that were not adequately considered some 40 years earlier when the city obtained its rights. The doctrine has been accepted in more limited contexts in other state courts and widely studied throughout the West as a vehicle for judicial reallocation, but its forceful application is confined largely to California.

The committee believes that the values the public trust doctrine seeks to protect are best accommodated in comprehensive water planning processes and through public interest review of new appropriations and transfers. Rather than a technique or option for dealing with the effects of transfers, the public trust doctrine should be seen as a remedial device available to the public when there has been a failure of the system to protect the public interest. The doctrine applies ad hoc and lacks precise standards to judge how water should be allocated and used by competing users. No after-the-fact remedy can deal precisely or effectively with resource use and allocation, so the most valuable function of the doctrine is to signal the need for processes to avoid its judicial application (Graff, 1986). The recommendations suggested throughout this report are intended to address all values comprehended by the public trust doctrine and thus avoid the need for its selective application by the courts.

Clean Water Act, Section 404

Almost any water diversion or storage facility involves filling wetlands and may require a Section 404 permit from the U.S. Army Corps of Engineers under the Clean Water Act. Because some type of structure or facility frequently is necessary for a water transfer to occur, there is a potential federal "handle" on transfer projects that opens the possibility of extensive, detailed federal reviews and conditions. Although the limits of Section 404 authority have not yet been defined, the federal government could use the 404 program to regulate the effects of transfers. To date, the federal government has not fully exercised its potential authority under the statute.

Where state policies exist and are not in conflict with federal law, they can be implemented through the federal permitting process. It is the policy of the Corps of Engineers to follow state policy wherever possible. Therefore, to the extent that state law deals with the effects of transfers, it increases the chances that the federal government will not become involved. But in the absence of state policy, the Corps of Engineers and the Environmental Protection Agency will set their own policy for the state in the course of making Section 404 permit decisions.

Ad Hoc Negotiation

Many of the diverse effects of transfers can be dealt with in negotiations between affected parties and the parties to the transfer. This can occur with or without the benefit of laws to prod cooperation, and it can include anyone the parties decide to admit to the process. The Windy Gap Project discussed in Chapter 6 is illustrative. Of course, parties with an arguable legal right under some statute and parties with access to legal mechanisms to delay or increase the costs of a project have the greatest bargaining power in such negotiations.

The fundamental problem with negotiated resolutions that are not required by water transfer laws is that they produce uneven and incomplete results. Some parties will be treated better than others; some may be overlooked. Some transfers will entail high public visibility and political interest to empower affected parties; others will not. A consistent, reasonably predictable approach applicable in all transfers of significant size or impact will be more equitable and will encourage more desirable transfers. This argues for a formal process informed by impact assessments, comprehensive plans, and public interest reviews of particular transfers. Negotiated resolution of the issues in such a setting should be superior to ad hoc negotiations that vary with the political and economic power of the various parties.

Other Legislation

Although the transfer of appropriative water rights has always been possible, it is only recently, in the era of full appropriation of many western streams, that reallocation has become the main source of water for new enterprises. States and their citizens are realizing that voluntary transfers among private parties may affect an array of interests that are not adequately protected by the laws and processes that govern transfers. Western states, the federal government, and water districts all have opportunities and responsibilities to deal with the effects of transfers. The most direct way to do so is through programs of policy planning, impact assessment, and public interest review of projects. In addition, several other laws can deal with the effects of transfers.

State instream flow laws are important for limiting the environmental and economic effects of transfers. The structure and administration of these laws may have to be modified so they can be integrated with transfer laws. Laws that treat instream flow rights like other appropriative water rights (as in Colorado) provide protection against the adverse effects of transfers through the normal operation of state water law and its no injury rule. But these laws can also be used more expansively. At minimum, wider purposes than fishery protection need to be served. Instream flow protection can be addressed by requiring a portion of salvaged water (as in Oregon and Washington) and other water that is the subject of a transfer to be dedicated to flow maintenance. This would be an exaction similar to requiring land developers to dedicate land for streets, open space, or schools. Federal project water should be used in ways that are compatible with state instream flow laws, and transfers should be allowed only when these state laws are satisfied. Ultimately, protection of instream flows will depend on the acquisition of senior rights and the use of transfers to shift away from consumptive to instream flow uses.

State water quality protection goals can be furthered in the transfer process if statutes and procedures are clarified to specify that purpose. Stricter controls of point source discharges, without the self-destructive provisions in some state laws that subordinate quality protection to an unfettered right to appropriate water, are needed. Nonpoint source pollution of water use should be examined and checked at the time a right is transferred. And limits can be placed on the quantity and manner of use if they degrade water quality.

Laws allowing and encouraging special water districts to market water beyond their boundaries are generally beneficial. There should be mechanisms, however, for ensuring that public objectives beyond water supply are served by these transfers. Federal legislation consenting to the leasing or other use of Indian reserved water rights outside reservations should include a means of reviewing the impacts on the reservation culture, economy, and ecology.

REFERENCES

Arizona Game and Fish Department v. Arizona State Land Department, 24 Ariz. App. 29, 535 P.2d 621 (1975).

Bonham v. Morgan, 102 Utah 2d, 788 P.2d 497 (1989).

City of El Paso v. Reynolds, 597 F. Supp. 674 (1984).

Crawford, S. 1990. Dancing for water. Journal of the West 32:265-266.

Getches, D., L. MacDonnell, and T. Rice. 1991. Controlling Water Use: The Unfinished Business of Water Quality Protection. Boulder: University of Colorado, Natural Resources Law Center.

Graff, T. 1986. Environmental quality, water marketing and the public trust: Can they coexist? UCLA Journal of Environmental Law and Policy 5:137.

In re Application of Sleeper, No. RA 84-53 (N.M. District Court for Rio Arriba County 1985).

MacDonnell, L. 1990. Transferring water uses in the West. Oklahoma Law Review 43:119.

National Audubon Society v. Superior Court, 189 Cal. Rptr. 346, 658 P.2d 704 (1983).

National Research Council (NRC). 1987. The Mono Basin Ecosystem: Effects of Changing Lake Level. Washington, D.C.: National Academy Press.

Roos-Collins, R. 1987. Voluntary conveyance of the right to receive a water supply from the United States Bureau of Reclamation. Ecology Law Quarterly 13(4):773-878.

Salt River Water Users' Association v. Kovacovich, 3 Ariz. App. 28, 411 P.2d 201 (1966).

Sax, J. 1990. The Constitution, Property Rights, and the Future of Water Law. Western Water Policy Project. Boulder: University of Colorado School of Law, Natural Resouces Law Center.

Shokal v. Dunn, 109 Idaho State Supreme Court 330, 707 P.2d 441 (1985).

Sporhase v. Nebraska, 458 U.S. 941 (1982).

Tarlock, A. D. 1990. State groundwater sovereignty after Sporhase: The case of the Hueco Bolson. Oklahoma Law Review 43:27-49.

United States v. California, 438 U.S. 645 (1978).

United States v. Winters, 207 U.S. 564 (1908).

Wahl, R. W. 1989. Markets for Federal Water: Subsidies, Property Rights, and the Bureau of Reclamation. Washington, D.C.: Resources for the Future.

Western Governors' Association Water Efficiency Task Force. 1986. B. Driver, ed., Western Water: Tuning the System. Denver: Western Governors' Association.

Western Governors' Association Water Efficiency Working Group. 1987. Water Efficiency: Opportunities for Action. Denver: Western Governors' Association.

Willey, Z., and T. Graff. 1988. Federal water policy in the United States—An agenda for economic and environmental reform. Columbia Journal of Environmental Law 13:325.

4

Assessing Water Transfers and Their Effects: An Introduction to the Case Studies

Captain Renaud: What in heaven's name brought you to Casablanca?
Rick: My health; I came to Casablanca for the waters.
Captain Renaud: The waters? What waters? We're in the desert.
Rick: I was misinformed.

From the film *Casablanca* (1942)

Proposals to alter the way water is allocated in the West must necessarily consider the network of interrelated impacts that such changes might have on public and private interests. The committee elected to study these effects by undertaking a series of case studies; seven areas were selected to reflect a diverse view of the region's experience with water transfers. It was hoped that this in-depth look at specific experiences would provide insights about how to evaluate water transfer proposals in general. In analyzing the case studies, both individually and in the aggregate, the committee attempted to develop and use a systematic evaluation strategy to ensure that the analyses were both thorough and consistent. This treatment of the effects of transfers is modeled on the "four-account" system for evaluation of water projects embodied in *Principles and Standards for Planning Water and Related Land Resources* (Water Resources Council, 1973, 1980).

The characterization of third party effects will evolve as transfer activity increases. Still, it is the committee's belief that the evaluation strategy outlined here reflects the existing consensus on the range of physical and socioeconomic concerns that must be represented. In addition, it gives attention to intangible concerns that, although difficult to assess, must be considered.

ELEMENTS OF THE EVALUATION SYSTEM

Seven locations investigated in this report were selected to highlight the main issues that arise when transfers are contemplated or

Planning for Water Resources: The Principles and Standards Approach

The question of how to measure the effects of water development and management activities is not new. During the 1960s the merits of federal water resource development projects were measured by conducting benefit-cost analyses that focused on economic impacts. During the past 25 years the federal government has taken the lead in trying to expand the range of impacts considered. The precise structure and weighing of the accounts have varied over time, but the underlying notion of a system of evaluation has endured.

The evaluation system was embodied in *Principles and Standards for Planning Water and Related Land Resources* (P&S) (Water Resources Council, 1973, 1980). The P&S were developed as a basis for evaluating plans involving federal or federally licensed actions. The scale of activity ranged from site-specific problems to multipurpose plans for river basins. Although not designed specifically to look at the effects of water transfers, these guidelines offer insights to help illuminate third party impacts.

The P&S approach offered a four-account system designed to represent the range of public interests and concerns that should be considered in water resource decisionmaking. The four accounts are national economic development (NED), regional economic development (RED), environmental quality (EQ), and social well-being (SWB) (later changed to other social effects (OSE)). The latter two categories may prove especially pertinent to the potential impacts of transfers.

The NED account is an economic efficiency criterion. Positive contributions to NED are increases in the economic value of the national output of goods and services; adverse effects include the opportunity costs foregone in committing resources to the proposed project or activity. Suggested methods of analysis include assessment of net benefits, cost-effectiveness, benefit-cost, and discounting of future benefits.

The RED account records potential effects on income and employment within the region to be affected by the activity. From a national efficiency perspective, RED is an inappropriate criterion because net economic effects favorable to a region may be adverse from the standpoint of the national economy. But since regional development effects have in fact been the bedrock of water development politics, it is useful to examine them when assessing the possibilities and constraints of innovation.

The EQ account "identifies beneficial and adverse effects on ecological, aesthetic, and cultural attributes of natural resources."

As surrogates for the economic parameters used in NED and RED accounts, other factors are used to measure EQ resources. These include "institutional recognition," "public recognition," environmental "customs and traditions," and "technical recognition," meaning the significance as judged based on scientific knowledge or judgment.

The OSE account includes three categories of particular interest in terms of water transfer discussions: community impacts, displacement, and long-term productivity. Community impacts include income and employment distribution with particular reference to minorities and low-income households.

SOURCE: Water Resources Council (1973, 1980).

conducted. Each site was visited either by the full committee or by an assigned subcommittee; the information gathered from the literature and from experts at each site included a brief history of the settlement and water use patterns in the area, a description of the physical and socioeconomic setting, an overview of actual and potential transfer-related activities in the area, and—most importantly—a characterization of the type of transfer and its impacts. The elements of the evaluation strategy fall into three broad categories: transfer characteristics, third party interests, and the nature of third party effects.

Transfer Characteristics

Transfers can be voluntary or involuntary and involve a number of different state and federal procedures. Water transfers vary in type and impacts—and do not necessarily always cause harm to third parties. For the evaluation system used in this report, there are five key types of transfers under which third party effects may arise:

1. Change in ownership of the water right. In some jurisdictions the change in ownership of a water right or contract entitlement to storage water from a federal reservoir may be treated formally as a transfer. However, a change in ownership alone does not usually lead to third party effects; as long as the water continues to be used for the same purpose at the same location, no third party effects need occur. For example, the purchase of farmland by a city in need of assured water supplies to support future development may or may not entail third party effects. In some cases, the city will hold the rights, but farming will continue, and no immediate impacts will be felt. In other cases, even though the water continues to be used for agriculture, land values in the nearby community may drop. At some future time, when the city decides to use the water acquired through the farmland purchase, other third party effects may arise. Thus, when the type of use and the place of use remain unchanged, transfers of ownership do not necessarily lead to third party effects, although they may.

2. Change in the point of diversion (for a state water right) or change in point of delivery (for federal project water). This type of transfer is fairly common and usually includes informal consideration of the impacts on other parties who hold water rights. A typical change in point of diversion might be caused by the need to relocate a water management structure such as a diversion dam or well. If the needed change is over a considerable distance, the likelihood of third party effects increases.

3. Changes in place, purpose, or period of use. Any change in place, purpose, or timing of water use, such as a change from irrigation use to municipal or industrial use, can entail impacts on third parties. This type of transfer is most commonly associated with market-like exchanges.

4. Change in system operations. This type of de facto transfer usually results from court actions or negotiated arrangements. Such transfers typically result from some modification in the way a project or water supply system is operated. Thus, for example, a reservoir project might be operated by closing the dam at the end of the irrigation season and reopening it only at the beginning of the next season. If, however, project operations are ultimately changed to accommodate migrating fish, and flows are maintained year-round as a consequence, the quantity of water stored at the dam will be reduced, and third party effects may result.

5. Out-of-basin diversions. Interbasin transfers—the diversion of water from one hydrologic basin to another—do not necessarily involve a transfer of water rights. For instance, in some cases, unappropriated water in one basin may simply be moved to another basin. In other cases, however, existing water rights are involved when water is moved. Both types of interbasin transfers may bring third party effects.

Third Party Interests

The parties or interests affected by water transfers as well as the nature of the effects themselves may be quite varied. For each case study, the committee identified the interests and parties affected and characterized the type of effect. For convenience, the interests and parties were grouped as follows:

1. **Other water rights holders**. Included in this category are the holders of other water rights whose water supplies could be diminished or qualitatively impaired by a transfer or whose flexibility in using water might be affected.

2. **Agriculture**. Third parties in this sector include other agricultural water users whose sources of supply may be either quantitatively or qualitatively affected or who are subject to changes (either positive or negative) in the cost of water as a consequence of the transfer.

3. **The environment**. Transfers have the potential to exert a broad range of impacts on the environment, both positive and negative. Transfers can affect instream flows needed for recreational uses, fish and wildlife habitat, and the potential to generate hydroelectric power. Changes in the structure and health of aquatic ecosystems generally, or with particular effects on endangered species, and changes in water quality that threaten human health or aquatic ecosystems can also result.

4. **Urban interests**. Third party impacts on the urban sector can include the loss or gain of economic activity; diminution of the tax base, as when water previously used in industrial or agricultural activity is transferred to a public entity, thereby eliminating the taxable private activity and replacing it with a tax-exempt government function; and the loss or gain of recreational opportunities.

5. **Ethnic communities and Indian tribes**. Affected parties in this sector include ethnic communities that are especially dependent on the availability of water and particular ways of using water; Indians and tribal communities, many of whose water rights have been clearly determined; and residents of other special agricultural and rural communities who may value agricultural and rural lifestyles that could be jeopardized by water transfers.

6. **Rural communities**. Transfers of water from rural areas can affect residents who are not parties to the transfer or water users as a consequence of erosion in the local tax base resulting from the decline or disappearance of water-based economic activity. These transfers can also have adverse (or, less commonly, beneficial) impacts on communities and businesses that are linked to water-using industries affected by a transfer. Thus, for example, firms that supply feed and fertilizer to irrigated agriculture in a given region could be hurt by large-scale transfers of water out of that region. Rural interests could

also be hurt by the loss of their natural resource base occasioned by such a transfer.

7. Federal taxpayers. Federal taxpayers could become third parties to transfers that either enhance or diminish the productivity of the national economy or confer windfall gains on the buyers or sellers at public expense. Similarly, federal taxpayers may be third parties when transfers have impacts on national social or environmental objectives.

Nature of Third Party Effects

Third party impacts can be divided into economic effects, environmental effects, and social effects. Economic effects include positive or negative contributions to the economic value of the region's or the nation's goods and services and adverse effects stemming from opportunities foregone because of the transfer of water. Economic impacts also include changes in income and employment, the distribution and composition of local or regional populations, and the fiscal condition of state and local government. Environmental impacts include all water-related environmental effects such as the impact of alterations in instream flows on fish and wildlife, or recreation, changes in water quality, and the implications of a water transfer for wetlands and riparian ecosystems. Social impacts include noneconomic (and difficult to measure) effects on the quality of rural communities and municipalities. Intangible impacts include changes in the quality of community life, feelings of "connectedness" to the land, and a sense of control over an area's destiny. The long-term productivity and security of the community are affected when natural resources such as critical watersheds, productive cropland and rangeland, and riparian and aquatic ecosystems are lost.

INTRODUCTION TO THE CASE STUDIES

To highlight the nature and impacts of water transfer activities, the committee examined seven case studies (see Figure 4.1):

- Truckee-Carson basins in Nevada (Chapter 5),
- Colorado Front Range (Chapter 6),
- northern New Mexico (Chapter 7),
- Yakima Valley in Washington (Chapter 8),
- central Arizona (Chapter 9),
- Central Valley in California (Chapter 10), and
- Imperial Valley in California (Chapter 11).

FIGURE 4.1 Areas studied by the committee.

These cases illustrate not only a diversity of participants and effects, but also greatly varying political and social environments. In examining the cases the committee used the evaluation system described earlier in this chapter and summarized in Table 4.1.

To gather information and stimulate discussion, the committee met informally at each case study site with representatives selected from a range of interests—agriculture, water agencies, urban planners, environmental groups, local tribal populations, and other minority interests as appropriate (see Appendix B). The committee encouraged its guests to participate in frank discussions about the real and potential effects of water transfers on the people of their states and regions, with the goal of developing an accurate impres-

TABLE 4.1 Factors to Consider When Assessing Potential Water Transfers

---

Type of Transfer

- Change in ownership
- Change in point of diversion
- Change in use
- Change in systems operation
- Out-of-basin diversion

Primary Process for Transfer

- Voluntary
- Involuntary

Primary Market Forces for Transfer

- Government
  - Local
  - State
    - Executive
    - Legislative
    - Judicial
  - Federal
    - Executive
    - Legislative
    - Judicial

Affected Parties

- Rural communities
  - Support services
  - Erosion of tax base
  - Loss of natural resource base
- Agriculture
  - Remaining water users
  - Reallocation of rights
- Ethnic communities and Indian tribes
  - Ethnic communities
  - Indian communities
  - Agricultural maintenance and expansion
  - Other
- Environment
  - Instream flows
    - Recreation uses
    - Fish and wildlife
    - Hydroelectric power
  - Water quality
    - Damages to water users
    - Human health
    - Ecosystem effects
  - Ecosystem protection
    - Endangered species
    - Wetlands
    - Riparian habitat
  - Estuaries
- Urban interests
  - Intrastate transfer constraints
  - Tax-exempt status changes
- Federal taxpayers
  - National economic concerns
  - Windfall profits
- Other water rights holders
  - Junior rights
  - Senior rights
  - Loss of flexibility

Nature of Effects

- Economic (national/regional)
  - Lost revenue
  - Lost opportunities
  - New revenue
- Environmental
  - Instream/fish and wildlife
  - Recreation
  - Water quality
  - Wetlands
- Social
  - Rural communities
  - Municipalities
  - Other

---

sion of the various scenarios. The committee supplemented its interviews with reviews of the appropriate literature and the expertise of individual committee members. The case study approach has both strengths and flaws. Its greatest strength is the honesty of the discussions; its main weakness is a necessary brevity and lack of depth.

For greater elaboration the committee refers readers to the sources cited in the chapters.

The committee considered two levels of analyses central to its assignment: (1) analyses of individual cases in order to develop a thorough understanding of the site-specific causes and consequences of water management actions and third party effects and (2) analyses of the entire body of case study information and other data sources to formulate and evaluate broad policy and programmatic recommendations addressed to the management of third party effects.

In undertaking the analyses of the individual case studies, the objective was not to judge the desirability of actions taken or not taken in a particular case. Rather, the committee used the analyses to help understand the causes and consequences of third party effects. Each third party effect was assessed in terms of its economic, environmental, and social impacts. In so doing, the committee recognized that water managers, political decisionmakers, and citizens will usually assign different weights to the three categories of impacts, depending on their own personal, institutional, and political perspectives. Thus the committee did not attempt to assign weights to different types of impacts or to otherwise rank them in importance.

In examining the entire body of case study information the committee sought to characterize the causes and range of third party effects, identify those that were pervasive and those that were unique, identify the incidence of third party effects, and understand the actions available to mitigate or remedy the effects. This information was then used to identify explicit program or policy areas for further analysis and to justify development of the recommendations given in Chapter 12. Where appropriate, supplemental sources of information and studies published in the scientific literature were used to aid the committee in arriving at its final recommendations.

CRITICAL ISSUES

The case studies were conducted to examine actual and potential water transfers in diverse areas of the American West to understand better the nature, scope, impacts, and institutional setting of water transfers. Every situation is unique, but there are important commonalities. In some places—for example, Nevada's Truckee and Carson basins, central Arizona, and the Front Range of Colorado—the committee found that substantial transfer activity has occurred, sometimes involving an actual change in type or location of water use and other times involving only the transfer of rights. In other places, such as the Central Valley of California, few transfers have occurred despite substantial pressures from prospective buyers. In yet other instances, such as the Yakima Valley of Washington State, transfers have been identified as a potential management option, but there are few incentives to stimulate them. The impacts of the transfers that have occurred also vary: the Imperial Valley case shows substantial benefits with few problems; potential problems were addressed through the lengthy negotiations that preceded the transfer agreement. Many cases, on the other hand, provide illustrations of unresolved third party impacts.

The problems and opportunities illustrated in the various case studies are not unique and will arise elsewhere throughout the West. Thus valuable lessons for future federal and state water allocation policy can be drawn from the studies. Three critical issues, in particular, received recurring attention during the course of this study: area-of-origin protection, instream uses, and transaction costs.

Area-of-Origin Protection

If water transfers are to be used as one mechanism of responsive water management, the equity issues related to area-of-origin impacts will require continued attention. California enacted area-of-origin protection laws in the 1930s partly in response to Los Angeles' controversial efforts to export water from the Owens Valley. But these and other early efforts to protect areas of origin were largely ineffectual. Until recently, when increased public awareness and insistence on protecting environmental values focused renewed attention on the potential area-of-origin impacts of interbasin transfers, there was little interest in strengthening these laws.

Impacts on areas of origin are addressed inadequately in most states. Although an individual farmer might benefit from selling water rights to satisfy growing urban demand, rural counties are left trying to protect their tax bases, environments, cultures, and economic futures. The Arizona Ground Water Management Act, as one example of this dilemma, allows water to be transferred from agricultural areas to municipal and industrial uses but has no mechanisms for considering the interests and values of the areas where the water originates.

In Colorado, the irrigation ditch companies and municipalities who are the primary parties to transactions moving water from Colorado's Arkansas River valley to growing cities appear satisfied with the outcome. Still, not all the affected interests were represented and considered. There were no procedures, other than the Colorado water court process, for weighing concerns about water quality or soil erosion, and only limited efforts were made to judge the extent of decline in economic activity or reduced support for community infrastructure that is likely to follow the transfers. The lack of mechanisms outside the court process has frustrated west slope mountain communities in Colorado, who believe that export of their water to the Colorado Front Range will ultimately dry up their present economies—economies now based primarily on water-dependent tourism and outdoor recreation. The communities believe that interbasin transfers foreclose their future options.

Subtler concerns are expressed by people in communities where social and cultural values may depend on protecting existing water uses and having water naturally in place. They fear that out-of-basin transfers will undermine the integrity of the communities themselves. In northern New Mexico, for example, the seemingly neutral institutions of adjudication—which can include a public interest review of new appropriations and transfers—do not fully reflect the concerns of the Hispanic community. Lifestyle, community organization, and personal relationships are all intimately related to traditional water uses and allocation arrangements. The state system, however, values water essentially as a commodity; it does not see water as an element of community cohesion, history, and collective aspirations.

Instream Uses

One of the major issues facing decisionmakers is the balancing of consumptive and nonconsumptive, or instream, uses. Prior appropriation allows private rights to be created in public resources, and throughout the history of the West appropriative water rights were typically held by private parties or public water providers for consumptive uses. (An exception was made for hydropower rights, which could be exercised only by building facilities in the stream to take advantage of flowing water.) However, many of today's demands are for nonconsumptive uses that require flowing water to be left in the stream. With few exceptions (e.g., New Mexico), most western states have enacted programs that legally recognize instream uses. Ordinarily, a public entity holds these rights. In most cases, instream flow rights are junior to existing uses.

Several of the case studies illustrate that transfers can be used to help satisfy demands for instream flows with more reliable senior rights. Existing institutions do not encourage transfers for nonconsumptive uses and may in fact inhibit them. The Truckee-Carson experience illustrates the positive potential of agriculture-to-wildlife and wetland transfers. Incentives for transfers in this case included both the threat of a judicially mandated involuntary transfer and the provision of funds to induce water rights sales. Judicial decisions reallocating water to Indian tribes and wildlife protection laws created strong incentives for transfers of water needed to maintain Pyramid Lake and Stillwater National Wildlife Refuge. Federal funds were authorized to facilitate the transactions. Thus, in the Truckee and Carson basins as in the Imperial Valley, transfers occurred because a court-ordered involuntary reallocation was possible and well-funded buyers were available.

Voluntary transfers to instream uses have not occurred in other areas, such as the Central Valley and the Yakima Valley, where courts have not intervened or where they have confirmed existing uses. The refusal of courts to expand Indian entitlement in the Yakima basin leaves the status quo firmly in place and provides few, if any, real incentives to support legislative efforts to encourage voluntary transfers.

Transaction Costs

Public policy has been greatly influenced by the concept of transaction costs. Transaction costs are the costs of negotiating and enforcing transfer agreements and clarifying property rights so that transactions may proceed. The higher the transaction costs, the lower are the profits or economic benefits that accrue to those transferring the water from one use to another, and therefore the lower are the incentives motivating transfers. Some transaction costs are a necessary part of enacting any water transfer. These include costs incurred when the range of parties engaged in the transfer process is expanded to include third parties. Other costs, however, result from ambiguous policies and criteria for transfer approval and from duplicative efforts to provide information about a transfer and its potential impacts.

The ability to impose transaction costs on those proposing to transfer water, an ability conferred by state laws governing who may effectively object to a transfer, represents bargaining power in the water allocation process. Market transactions are undertaken for economic gain, based on the perception that water supplies will generate higher returns in their new use. The power to erode this expected gain by imposing transaction costs gives third parties leverage in the water reallocation process.

In general, the effects brought by transaction costs can be beneficial when the costs are incurred in empowering and accommodating traditionally underrepresented third party interests. In other words, society agrees to incur costs beyond those basic to the process of buying and selling water in hopes that its values will be more broadly represented and the hidden costs of transfers addressed. As interest in water transfers as a voluntary reallocation tool has increased, there has been a growing call to "streamline" the process so that market transfers are encouraged. The case studies suggest, however, that this goal must be tempered by the reality that some increased transaction costs are necessary if we are to address third party effects adequately. That does not mean there are not opportunities to improve the transfer procedures in use in the West and reduce some costs. However, transferring water is no simple matter; thus no simple and inexpensive process will be able to meet the needs of buyers, sellers, governing bodies, and affected third parties equitably.

REFERENCES

Water Resources Council. 1973. Principles and Standards for Planning Water and Related Land Resources. Washington, D.C.: U.S. Government Printing Office.

Water Resources Council. 1980. Principles and Standards for Planning Water and Related Land Resources. Washington, D.C.: U.S. Government Printing Office.

5

The Truckee-Carson Basins in Nevada: Indian Tribes and Wildlife Concerns Shape a Reallocation Strategy

The arid lands of the Truckee-Carson basins of western Nevada are in transition: urbanization is increasing and the agricultural base is declining. Significant amounts of land are being allocated to the preservation of Indian cultures and wildlife habitats. The area's transition mirrors the land use changes that are occurring in many areas and represents another stage in the constant redefinition of the West (Athearn, 1986).

The Truckee-Carson transition is characterized by creative use of water transfers to solve the intense water allocation conflicts that have arisen between traditional and nontraditional users. Water use conflicts are being resolved through a combination of litigation, legislation, voluntary transfers, and consensus-building processes. Litigation was used successfully by a relatively powerless minority, an Indian tribe, to change the balance of power among the major water users in the basin. This new balance provided the incentives for most major water users to address basinwide problems comprehensively and to devise creative solutions that should promote both efficiency and equity. In 1990, Congress played a critical role in the long trek from conflict to consensus. The resulting legislation (the Truckee-Carson-Pyramid Lake Water Settlement Act, Title II of P.L. 101-618) may prove to be a model for other western basin settlement acts.

In the past decade, tribal people have been able to gain unprecedented power to influence the water allocation process through federal environmental and tribal laws, a power denied them in earlier adjudications. Ironically, the Indians' success, along with the ongoing effects of a prolonged drought and of substantial urban and agricultural diversions of water, has contributed to the reduction of a formerly thriving wildlife habitat—the Stillwater Marsh and surrounding Lahontan Valley wetlands. These wet areas have shrunk from some 40,000 acres to a present size of less than 6,000 acres. Wetlands in the Lahontan Valley have been variously affected by the advent of irrigated agriculture: first, by diversion of the historic Carson River flow; then by overdiversion of the Truckee River flow; and, most recently, by efforts to improve the efficiency of the Newlands Irrigation Project. These remnant wetlands were able to coexist precariously with an upstream irrigation district but not with the additional goal of restoring the native fishery. To the extent that the tribe's right to maintain inflows to Pyramid Lake and its fishery is met, there is a reduction in the amount of water available to the irrigation project and, with disproportionate impact, the wetlands.

Water transfers from irrigation to wetland maintenance hold promise for stabilizing the wetland area, increasing the Newlands Irrigation Project's efficiency, and allowing the irrigation economy created in the first half of this century to adjust to the fundamental shift in water resource values that has occurred in the past two decades. The changes occurring in the Truckee-Carson basins suggest that management to promote multiple uses of water and protection of biodiversity is increasingly important. Restoration to the maximum extent possible—including protection of historic ecological diversity and Indian cultural values—is gradually replacing the past emphasis on consumptive water uses.

THE SETTING

Rainfall averages about 9 in. (230 mm) per year throughout the lower Truckee-Carson basins, and the available surface supplies are fully appropriated by adjudications begun in the reclamation era. There are substantial reallocation pressures as urban areas grow, irrigated agriculture continues to decline, and environmental values and the importance of an ecological balance become recognized. This area is a microcosm of the reallocation patterns and institutional changes that are emerging in many areas across the western United States. Both traditional and nontraditional users claim substantial shares of the area's water resources. In addition to the usual conflicts between municipal and industrial uses and irrigated agriculture (both non-Indian and Indian irrigators), there is competition from aboriginal

Indian fisheries, wetland maintenance, and migratory waterfowl production for some 750,000 acre-feet (925,000 megaliters (ML)) per year of decreed rights.

Four major water-using interests compete for the basin's limited supplies (Truckee meadows urban area, Newlands Project agriculture, Pyramid Lake endangered fish species, and the Lahontan Valley wetlands). All users depend primarily on two rivers, the Carson and the Truckee, which drain from the eastern slope of the Sierra Nevada Mountains in California into two closed Great Basin systems, Pyramid Lake and the Lahontan Valley marshes, and the Stillwater, Carson sink, and Carson Lake marshes (Figure 5.1). Ground water re-

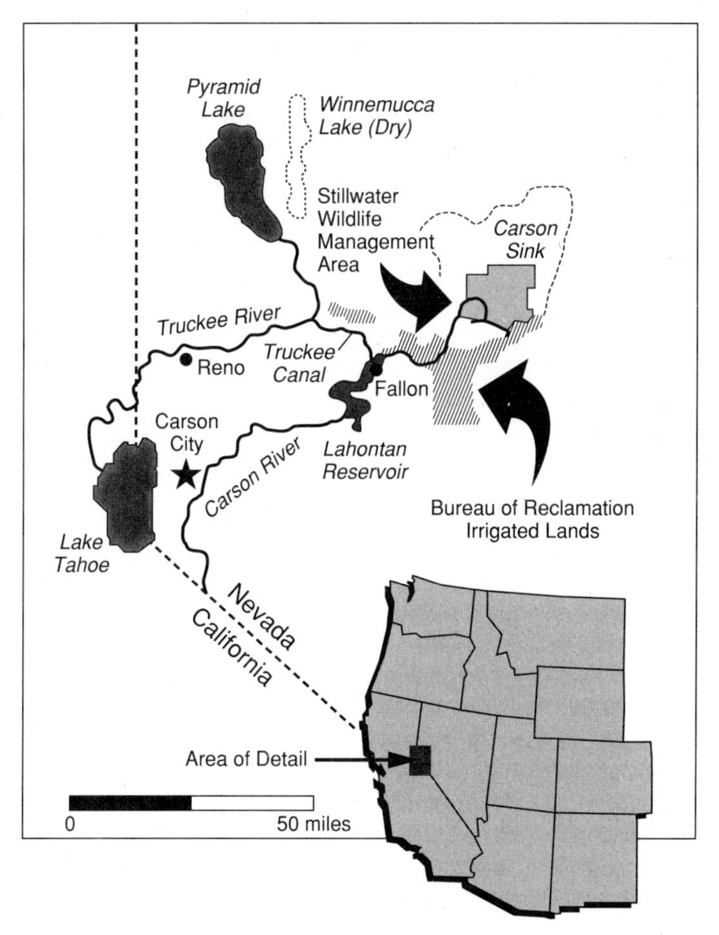

FIGURE 5.1 Truckee-Carson river basins in western Nevada.

sources are managed on a safe yield basis and can provide only about 20 percent of the supply in the urban area. The rapidly growing metropolitan Truckee meadows area (the towns of Reno and Sparks) dominates the upper Truckee River water use, and 75 percent of the original agricultural uses to the west of Reno have already been converted to municipal and industrial uses. Irrigated agriculture still dominates the use of water in the lower Truckee and Carson rivers, although irrigation in the area has always been marginal in comparison with other areas of the West.

The Truckee basin was the home of one of the chief proponents of a western reclamation policy, Senator Francis G. Newlands of Nevada. Newlands, his fortune in part inherited from his legendary father-in-law, William Sharon, one of the developers of the Comstock lode, came to Nevada in the mid-1880s to manage the family holdings after the Comstock was exhausted and the state began to lose population. Newlands believed that agricultural development was key to the state's future (Truckee River Atlas, 1991), and he privately financed the Truckee-Carson Project in 1888. His project "was one of the most ambitious reclamation efforts of its day and it failed . . ." (Reisner, 1986). After he became a senator, Newlands, an admirer of John Wesley Powell, became the chief proponent of federal support for reclamation as the only way to stem the state's population decline (Hays, 1959). He drafted the Reclamation Act of 1902, and the Newlands Project was one of the first projects authorized after the passage of the act. The name Newlands was given not because of the "new lands" brought into production but to honor the senator.

The Truckee-Carson Irrigation District (TCID) is the Bureau of Reclamation's contract operator for the Newlands Project. Water rights on the project are held by the individual landowners, not TCID. A total of 232,000 acres was originally proposed for the project. The U.S. Department of the Interior (DOI) estimated that only 63,100 acres were irrigated in 1985 and that only 56,400 of these have project water rights. Still, the Bureau of Reclamation estimates that the irrigated area may increase because of Indian claims (U.S. DOI, 1986). Among the adverse impacts of the Newlands Project were the drying up of Winnemucca Lake, a national wildlife refuge adjacent to Pyramid Lake, a nearly 80-ft (24-m) drop in Pyramid Lake, and the near extinction of an Indian fishery.

Two Indian reservations assert potentially incompatible claims for the Truckee River. The Pyramid Lake Tribe claims additional flows to improve the quality, quantity, and timing of flows into Pyramid Lake, which in the words of the great western explorer John C. Fremont is "set like a gem in the mountains" (Truckee River Atlas,

1991). On average, 600,000 acre-feet (740,000 ML) flowed into the lake each year before man began to divert and consume the Truckee for irrigated agriculture. Now annual average inflows are reduced by about one-half of historic flows.

The additional flows are needed to allow the cui-ui—a large omnivorous sucker found only in Pyramid Lake, which the tribe considers sacred (Knack and Stewart, 1984)—and the Lahontan cutthroat trout to spawn. The cui-ui are listed as a federal endangered species, and the cutthroat are listed as a threatened species. Typically, the cui-ui migrate upstream from the lake to spawn in the lower Truckee River and then migrate quickly back into the lake. But spawning migration has been blocked or impeded by a delta that has formed in this century as a result of reduced river flow and a consequent drop in Pyramid Lake.

The Fallon Indian Reservation is on the eastern edge of the Newlands Project. This forgotten tribe was promised irrigation water in return for surrendering most of its lands when the Newlands Project was formed. As is often the case with respect to Indian irrigation, the promise went unredeemed for decades (Hurt, 1987). The tribe claims about 18,700 acre-feet (23,100 ML) per year to irrigate 3,100 acres. In 1990 the tribe achieved a considerable measure of justice. As part of the Truckee-Carson-Pyramid Lake Water Settlement Act (P.L. 101-618), a tribal settlement fund was created. Monies may be used to purchase up to 8,453 acre-feet (10,427 ML) per year, as well as for improvement of the existing irrigation system; total tribal water use is limited to 10,568 acre-feet (13,036 ML) per year.

Wetland ecosystem protection has been identified as a major water use in this already stressed area. The Carson River drains into the Lahontan marshes and the Carson sink. Before human settlement, some 85,000 acres of wetlands were sustained by the Carson River, and the Lahontan and Pyramid ecosystems were hydrologically distinct and harmoniously maintained. The development of the Newlands Project created a conflict between the two ecosystems that continues to the present.

In 1905, Derby Dam was constructed on the Truckee near Fernley, and the Truckee Canal diverted more than one-half of the Truckee's flow into the Carson basin. The diversion placed stresses on Pyramid and Winnemucca lakes. Diversions from the Truckee continued relatively unimpaired until the Pyramid Lake Tribe filed suit. In 1966, the Bureau of Reclamation began to adopt criteria to improve the operating efficiency of the project, and a federal court directed the Bureau of Reclamation to develop more stringent criteria for diversions from the Truckee River and operations of the Newlands Project. Wetland acreage began to diminish in the late 1960s, directly after TCID halted winter diversions solely for power generation. Starting in the 1970s, inflows into the Stillwater National Wildlife Management Area and other wetlands—which are home to bald eagles, feed American white pelicans, and support a wide variety of waterfowl—were diminished and became polluted. Reductions in the district's water conveyance efficiency have resulted in additional decreases. These factors, in conjunction with extended drought and other upstream uses, reduced the marshes to somewhere between 4,000 and 6,000 acres. To relieve this stressed ecosystem, wildlife interests are turning to the market to obtain rights denied them by early adjudications and subsequent reallocations.

Both the Truckee and the Carson are interstate rivers shared by Nevada and California, but there was no formal comprehensive allocation between the two states by Supreme Court decree, compacts, or acts of Congress until 1990. As part of the Truckee-Carson water rights settlement, Congress ratified the long history of judicial decrees, federal agreements, and unapproved compacts to provide a de facto allocation.

THE WATER DELIVERY SYSTEM

Carryover storage and transbasin diversions both play a large role in meeting existing water needs, but water diversions play the most significant role in the Truckee-Carson basins. The Carson River has been dammed only at the lower end to supply the TCID, and above Lahontan Reservoir its flows have great seasonal variation. To guarantee firm yield to the Newlands Project, the Truckee River is diverted to the Carson through the Truckee Canal, which runs from Derby Dam to Lahontan Reservoir. Several upstream reservoirs have been constructed on the Truckee, but there is insufficient carryover storage to provide normal flows during a 2- or 3-year drought cycle. Reno is investigating the extraction of ground water from Nevada's side of the Honey Lake interstate aquifer, about 75 mi north of the area. California ranchers have expressed concerns that the project will dewater the aquifer, but Nevada claims that its safe yield policies will prevent this from happening.

Lake Tahoe, which straddles the California-Nevada border, is both a natural lake of great beauty and a storage reservoir for the Truckee. Lake Tahoe could provide all the carryover storage that the area would need for the long term, but most of the water has been dedicated to in-place, nonconsumptive uses. Tahoe has a total capacity of about 122,160,000 acre-feet (150,685,000 ML), but the lake level is regulated to fluctuate a maximum of 6 ft, providing only 744,600 acre-feet (918,000 ML) of usable storage capacity. The six other small storage reservoirs constructed on the Truckee and its tributaries have a total capacity of 316,770 acre-feet (390,000 ML). Stampede Reservoir, built in the 1960s, contains the bulk of this capacity with 226,500 acre-feet (279,400 ML).

HOW WATER LAW HAS DEFINED RIGHTS AND CONSTRAINED REALLOCATION IN THE TRUCKEE-CARSON BASINS

The Initial Allocation

The law of prior appropriation has served to create firm property rights in the West's variable streams, and this function has played a dominant role in allocating the waters of these two rivers. Prior appropriation continues to operate as a constraint on reallocation options, but it also serves as a source of marketable rights for new traditional and nontraditional uses. The waters of the Truckee and Carson rivers were appropriated in the late nineteenth and early twentieth centuries, and this allocation remained relatively constant until the Indians were able to challenge it in the 1970s. Initially, allocation of the Truckee was determined by a series of agreements that fixed the flow rates in the Truckee to maximize the beneficial use of the river for hydroelectric power generation. Subsequently, the available supply of the Truckee River was adjudicated, as was the Carson River. The first dam at the mouth of the Truckee at Lake Tahoe was built in 1870; Truckee River flows, referred to as the Floristan rates, were first established in 1908. In 1915 the Bureau of Reclamation acquired the Lake Tahoe Dam in a consent decree that settled a condemnation suit. This decree gave the United States the right to raise the level of Lake Tahoe and obligated it to maintain the Floristan rates. Adjudication of the Truckee began in 1913, shortly after the California Conservation Commission recommended that the state seek an equitable apportionment of the Truckee and Lake Tahoe.

Truckee River water use is controlled by the *Orr Ditch* decree (*United States of America* v. *Orr Water Ditch Company*, 1944), the establishment of minimum levels for Lake Tahoe, and the Truckee River Agreement of 1935. The decree was not made final until 1944 and was not protected against collateral attack by the Pyramid Lake Tribe until 1983. This decree is administered by a federal water master; his basic task is to maintain a minimum flow at the California-Nevada state line relative to the level of Lake Tahoe.

The Carson River use is allocated by the *Alpine* decree (*United*

States v. *Alpine Land and Reservoir Co.,* 1980). It defines the water rights differently. The *Orr Ditch* decree defines rights as claimed appropriations, whereas the *Alpine* decree defines rights in terms of maximum consumptive use. Lake Tahoe storage is limited to 744,000 acre-feet (918,000 ML), and withdrawals from the storage pool are controlled by the Truckee River Agreement, in which DOI modified the original Floristan rates to allow additional flood storage. Minimum flows of 400 cubic feet per second (cfs; 11.4 m^3/s) are required during the winter months and 500 cfs (14.2 m^3/s) during the summer months. Lower winter flows are allowed when the level of the lake is between 6,226.0 and 6,225.25 ft (1,897.68 and 1,897.46 m).

Principal Interests in Water Reallocation Through Transfers

TRIBAL INTERESTS AND ENDANGERED SPECIES IN PYRAMID LAKE

The *Orr Ditch* decree supported the Newlands Project at the expense of the Pyramid Lake Tribe. In the 1960s the tribe began to challenge the original decree. Although the U.S. Supreme Court ultimately affirmed the finality of the decree, the tribe's objections have rendered the project's rights more uncertain in the past two decades. The Pyramid Lake Tribe's objection to the *Orr Ditch* decree is that, although it received 14,742 acre-feet (18,180 ML) for irrigation with the first priority, the tribe received no water for maintenance of the lake for traditional subsistence fishing and spiritual culture. Its way of life became imperiled after the Truckee Canal diverted about one-half of the virgin flow of the Truckee River to the Carson basin, causing the lake level to drop precipitously. After a long period of protest with the Bureau of Indian Affairs, the tribe turned to litigation to restore the Truckee flows. The tribe pursued three strategies: (1) modify the operation of the Newlands Project, (2) reopen the *Orr Ditch* decree, and (3) use the Endangered Species Act to control unallocated blocks of water.

The first and third strategies were successful. The Newlands Project controls the major available pool of water open to reallocation. Both the *Orr Ditch* and the *Alpine* decrees provide for generous water duties. Allowable deliveries for alfalfa are 4.5 acre-feet (5.5 ML) per year for bench lands and 3.5 acre-feet (4.3 ML) per year for bottom lands. In a major decision, a federal district court held in 1973 that the tribe was owed a trust duty to maintain the level of the lake and ordered DOI to modify the operation of the Newlands Project (*Pyramid Lake Paiute Tribe of Indians* v. *Morton,* 1973).

In response to this decision, DOI has issued progressively more stringent sets of Operating Criteria and Procedures (OCAP) for the Newlands Project, which assume that annual diversions from all sources to satisfy project irrigation rights will be reduced from 370,000 to 320,000 acre-feet (456,400 to 394,720 ML). The OCAP "are predicated on water being used on . . . water righted land in a similar manner as in the past with the project operating at reasonable efficiency" by reducing seepage, evaporation, and spill losses. Basically, the OCAP try to ensure that headgate water deliveries match court-decreed water duties; they establish efficiency targets for the project's distribution system and maximum allowable diversion that reduces annual diversions as the project's physical efficiency increases. If the Newlands Project's actual delivery efficiency exceeds the target efficiency for a water year, the district will receive a Lahontan storage credit for the Carson portion of two-thirds of the saved water. The net results are that more water should flow into Pyramid Lake and less into the Stillwater National Wildlife Refuge Area and that the Fallon Reservation still has an unsatisfied irrigation right.

The reductions required by OCAP only slightly ameliorate the drop of Pyramid Lake level. The tribe tried to reopen the *Orr Ditch* decree on the grounds that DOI's representation of both the tribe and the Newlands Project landowners in the adjudication constituted a conflict of interest. The U.S. Supreme Court held that the tribe had been adequately represented and thus the decree was final (*Nevada* v. *United States*, 1983). The tribe continues to object to other aspects of the *Orr Ditch* and *Alpine* decrees with limited success. For example, the tribe raised a public interest objection to the TCID's transfers of water from unirrigated lands with water rights to irrigated lands without water rights, but the Supreme Court dismissed it as simply a collateral attack on the *Orr Ditch* decree. The eligibility of lands to receive project water is still being litigated (*United States* v. *Alpine Land and Reservoir Co.*, 1989).

Despite these setbacks, the tribe has been able to assert its traditional water uses to achieve results through the Endangered Species Act. After the cui-ui was listed as endangered and the Lahontan cutthroat as threatened, the Ninth Circuit Court of Appeals held that Stampede Reservoir, built in the 1960s to supplement Reno's water supply, had to be managed to maintain the fish species instead of for municipal and industrial uses. Stampede storage can be released to TCID as part of the reservoir credit established by the yearly OCAP. TCID is entitled to a storage credit for the difference between the actual amount saved and the target amount. Further litigation has established that DOI need not carry over the credit from year to year (*Pyramid Lake Tribe of Indians* v. *Hodel*, 1989).

This decision has given the Pyramid Lake Tribe great leverage with upstream users, to complement the leverage that the earlier decision gave it with the Newlands Project. Reno-Sparks was unable to use Stampede Reservoir as a drought reserve and was forced to negotiate a legislative settlement. Under the Truckee-Carson-Pyramid Lake Water Settlement Act (P.L. 101-618), the Pyramid Lake Tribe, Reno-Sparks, the state of Nevada, and the United States reached a preliminary settlement agreement that requires the Secretary of the Interior to operate the Truckee River reservoirs to implement an earlier agreement among the parties. The reservoirs will continue to be used to maintain spawning flows, but Reno-Sparks obtained the right to use 7,500 acre-feet (9,250 ML) of Stampede water in "worse than critical drought conditions."

URBAN GROWTH IN THE TRUCKEE MEADOWS AREA

Litigation has enabled the Pyramid Lake Tribe to influence water use throughout the Truckee-Carson system. This creates substantial problems for water suppliers and developers in the Truckee meadows area; they have neither secure access to federal reservoirs nor clear access to alternative supplies. Interstate ground water use remains uncertain. To accommodate the continued unchecked growth in the Reno-Sparks area, water suppliers and developers have had to acquire existing irrigation rights. As a result of tribal pressure, the state legislature reversed its historic no-water-metering policy, but the community has yet to vote on the issue. The price of water rights has risen over the past 15 years from less than $750 per acre-foot to between $2,500 and $3,000 per acre-foot (less than $608 per ML to between $2,030 and $2,430 per ML).

Most of the existing water rights have now been acquired by WESTPAC, the water supply division of Sierra Pacific Power. The major potential third party effects of these reallocations will be on downstream users. Downstream water rights holders are entitled to return flows. Water used by Reno historically has been returned to the stream as treated effluent. The problem is that these discharges currently violate the Clean Water Act, and Reno and Sparks are seriously considering land disposal as a treatment alternative. This, if it occurs, may reduce return flows. All western states have struggled with how to allocate the entitlements to treated sewage. The Arizona Supreme Court, for example, has held that sewage effluent is wastewater; thus a city has no obligation to downstream users to continue to discharge it or to provide substitute supplies. The Nevada state engineer has taken the opposite position. More generally, the third

Urbanization is increasing in the arid Truckee-Carson basin, and new development such as these homes in Reno, Nevada, puts added pressure on the region's limited water resources. The area's transition away from an agricultural economy mirrors land use changes in many areas of the West. CREDIT: Todd Sargent, University of Arizona.

party and ecosystem effects of the use of treated sewage are not well understood.

Facilitated by the pressure and interest of Nevada's U.S. Senator Frank Reid, in 1990 the major parties in the basins negotiated a settlement to the ongoing dispute. Reno-Sparks may switch to land disposal of sewage to avoid the higher costs of building a treatment plant to comply with Clean Water Act standards; the Pyramid Lake Tribe wants an assured supply of more and cleaner water to maintain and raise the lake level; and federal and state officials have shown a willingness to consider the use of treated effluent for the Fernley wetlands. This solution may require the purchase of additional irrigation rights as replacement water for the tribe. The Newlands Project is the most likely source because prices are much lower than in the Truckee meadows. The parties are considering piping the effluent to a wetland site north of Fernley or to Dodge Flat on the Pyramid Lake

Indian Reservation. It will be spread to ensure rapid infiltration. Whether this will work and will satisfy tribal concerns remains unclear. The Dodge Flat alternative would have several advantages that could benefit Pyramid Lake fish species: water would remain within the Truckee basin, pumping could be timed to meet seasonal water needs in the river, and water quality would improve.

Sierra Pacific Power and the Pyramid Lake Tribe also negotiated a preliminary settlement agreement that was included in the final congressional legislation in 1990. With respect to the Truckee River, the agreement allows the management of Stampede Reservoir for spring spawning flows for the cui-ui and for drought reserve storage for the benefit of Reno-Sparks. Downstream users will be protected, although continued discharges from the Reno-Sparks sewage treatment plant may be necessary to firm up downstream rights. The tribe's role was twofold: to bring the pressure needed to provoke a settlement and to take the initiative to lead the parties to optimal solutions.

WETLAND ECOSYSTEM MAINTENANCE IN THE LAHONTAN VALLEY

Fish and wildlife protection is difficult to achieve in the area because the history of diversions has led to a situation where wetland ecosystem maintenance is last on the list of protected uses. Upstream consumptive uses have diminished flows into Stillwater National Wildlife Management Area, and efforts to restore the levels of Pyramid Lake have further limited flows to Stillwater. The wetlands have shrunk drastically because of a recent drought and continued upstream diversions. If this wetland area is to be preserved as a wildlife sanctuary, it will require more and better-quality water. Approximately 50,000 acre-feet (62,000 ML) of acquired water rights are needed to support a permanent 25,000-acre marsh. Furthermore, this must be "prime water" (i.e., nondrain water) in addition to the drain water that already is being received. This water will not come from the Truckee River; such reallocation is precluded because of the rights of the Pyramid Lake Tribe under Indian law and because of the Endangered Species Act and the demands of the Reno-Sparks area. Therefore the water must come from the Newlands Project. Since the Pyramid Lake Tribe is the beneficiary of all savings achieved through the OCAP, rights for the Stillwater refuge must come from reallocation of the Carson basin supplies.

The OCAP seek to serve all existing rights holders through more efficient operation of the project. Since 1987 the Environmental Defense Fund, consistent with its efforts to solve water problems through market transfers of water, has urged the use of voluntary water rights acquisitions to provide the necessary water for the management area. Environmental interests are proceeding on the assumption that irrigation use must be reduced by 50,000 acre-feet (61,680 ML) to restore the area. This water could come from increasing conveyance efficiencies over and above OCAP target levels or by retiring approximately 23 percent of the existing irrigated land (Yardas, 1987). In December 1989 The Nature Conservancy paid $135,000 to purchase water rights (about 400 acre-feet, or 490 ML) from 150 acres of marginal farmland. Under the Truckee-Carson-Pyramid Lake Water Settlement Act (P.L. 101-618), water rights may be purchased from willing sellers, but the Secretary of the Interior may target "purchases in areas deemed by the Secretary to be the most beneficial" for the purchase program. By the end of 1993 the Secretary of the Interior must conduct a study of the social, economic, and environmental effects of the water rights purchase program.

RECENT AND PLANNED TRANSFERS

Nontraditional water-using interests are behind three of the four major water transfers occurring in the Truckee-Carson basins. The major water transfers taking place in this basin are as follows:

1. **Newlands Project agricultural water to Stillwater National Wildlife Management Area for wetland maintenance.** This is a private transaction and produces satisfaction for most parties with no significant external effects.
2. **Newlands Project agricultural water to the Truckee meadows for urban use.** This is a voluntary transfer involving changes in ownership use within the basin. Some potential externalities from disposal of municipal effluent on land versus other direct uses are involved.
3. **Newlands Project agricultural water to Pyramid Lake Tribe for cui-ui fish species protection.** This reallocation involved an involuntary transfer effected through a change in system operation by the Bureau of Reclamation. This then led to "voluntary" increases of on-farm irrigation efficiency. This reallocation was forced by legal and political pressures to mitigate prior inequities. Had externalities been greater, some forms of compensation may have been required.
4. **Reno urban water from Stampede Reservoir storage to Pyramid Lake for cui-ui species protection.** This involuntary transfer was brought about by litigation resulting in a change in use of the water stored in Stampede Reservoir (originally proposed for provid-

ing drought contingency for the municipal users) and dedicating this water for delivery downstream during the spawning season of the cui-ui. This involved fundamental conflicts of social values and interests and was resolvable only through the courts in the absence of a market of willing sellers.

Voluntary water transfers seem an attractive way to address the third party effects of past water allocation choices because they can provide an accurate measure of the value of water and do not constitute a taking of existing rights. As clean water reaches the refuge through this process, the transfers represent an efficient and fair allocation of resources. This is especially true for marginal agricultural areas as in the Newlands Project. Crop production has not been central to the economic prosperity of the Fallon area because of the expansion of a naval air station. However, the solution to a complex ecological and social problem such as this will not be left to an unregulated market. To balance efficiency, equity, environmental, and local economic concerns, the water transfers must be consistent with criteria that seek to achieve these objectives, and the transfers must be monitored carefully.

There are two major problems with the acquisition alternative, although they do not arise in every transfer. When a buyer has identified a willing seller, it is assumed that the seller's water rights meet the buyer's needs. This assumption cannot be made when there are one buyer and a large number of willing sellers with rights of unequal quality. Water rights must be screened for eligibility.

The two problems center on eligibility and the assurance of "wet water." First, there are serious issues of implementation. Not every water right should be classified as eligible for acquisition. In the Newlands Project, not all lands have project water rights, and many rights are paper rights (i.e., the right is based on decreed amounts rather than actual beneficial use). Second, a water right that is not put to beneficial use may be lost by either forfeiture or abandonment. In Nevada, a pre-1913 right can be terminated only through abandonment, whereas a post-1913 right may be terminated more easily through forfeiture (*In re Manse Spring and Its Tributaries*, 1940). Because most water rights in Nevada predate 1913, they are difficult to terminate. Thus there is a need to ensure that the water rights that are purchased in fact represent "wet water" that has been put to active use for some period before the transfer.

Market price provides an indicator of the value of water, but it is not always the most reliable indicator. A variety of intangible or unquantified social and environmental values may be associated with the importance that an entire community attaches to a stable pattern of water allocation. There may be a community interest in the status quo that is not reflected in individual bargains. And even though the individual farmers, most of whom do not earn their livelihood exclusively from the irrigated land, may be satisfied with the compensation, such marketplace transfers do not prevent the character of the local area from changing. Changes may cause economic dislocations among those who depend on the agricultural base of the area. Some of these effects can be mitigated by social regulation of the market process. For instance, transfer deliveries can be delayed to allow suitable replacement crops; yearly and total ceilings can be placed on the amount of water that may be acquired for a given use; and irrigated parcels can be selected for acquisition with an eye toward minimum disruption of the area. Community participation mechanisms can increase the legitimacy of transfers.

Transfers began occurring in late 1989. The federal government has already appropriated money for water rights acquisition for Stillwater refuge. The Nature Conservancy purchased the permanent water rights on 150 acres from a farmer serviced by the TCID for conversion to instream flows. Wildlife management is a beneficial use under Nevada law (*State* v. *Morros*, 1988), and after a negotiation with the district, the Pyramid Lake Paiute Tribe, and the U.S. Fish and Wildlife Service, the state engineer approved the permanent transfer of 162 acre-feet (200 ML) of these rights to the Stillwater National Wildlife Management Area. Nevada law allows the transfer of water from irrigation districts if the proposed change of use does not adversely affect the cost of water to other water rights holders (Nevada Revised Statutes § 533.370). The TCID's possible objections were allayed by The Nature Conservancy's promise, which has now been assumed by the federal government, to assume the farmer's 40-year repayment obligation, and the tribe was satisfied by assurances that Truckee flows would not be altered. Also, 27 acre-feet (33 ML) were purchased by the Nevada Waterfowl Association and transferred under a Nevada statute that allows 1-year transfers on an expedited basis (Nevada Revised Statutes § 533.345). There is still a reluctance on the part of some farmers to sell water rights, but the price incentive is a powerful way to overcome resistance by creating willing sellers.

CONCLUSIONS

The Truckee-Carson area provides a graphic example of the complexities of water allocation in the West. This hydrologic system clearly illustrates the conflicts that can occur among urban, agricultural, environmental, and tribal interests, and it raises issues about federal water project management, endangered species, wetland protection, and water quality degradation. The many pressures for reallocation of water highlight the need for institutions and decisionmaking processes capable of responding efficiently and equitably to accommodate new and traditional water values in the region.

Over the past decade, various judicial decisions, administrative actions, and negotiated settlements have resolved some of the basin's allocation problems, but great controversy continues over many issues. The most promising solution resulting from recent negotiations is the passage of federal legislation authorizing and appropriating funds for voluntary acquisition of water from willing sellers in the Newlands Project for the benefit of the Stillwater National Wildlife Management Area. Although major issues regarding the nature and quantification of water rights "appropriate for transfer" still exist, some precedent-setting transfers have occurred and are gaining support among a wide array of interest groups and government agencies.

Substantial additional progress toward meeting both environmental restoration and tribal equity goals could be achieved, while protecting affected agricultural communities, if water rights status could be ascertained quickly and finally. With sufficient federal, state, and private funding, all parties to the basin conflicts could see their interests protected—provided they continue to bargain in good faith, respect the interests of their adversaries, adopt an integrated water management perspective, and diligently apply water conservation measures to consumptive uses.

The efforts under way to balance existing and new uses of water through negotiation and market transfers were stimulated by judicial decisions that reduced preexisting entitlements. The net effect of the court's findings in favor of the tribe's effort to preserve its fishery was to deprive Reno of a block of reservoir storage that it thought was reserved for future growth and droughts and to force the TCID to operate the Newlands Projects more efficiently. Once the tribe was so empowered, the Bureau of Reclamation had to adjust its management of Stampede Reservoir and delivery of water to the TCID. The tribe subsequently was included in broader negotiations about the Truckee's future.

Many of the issues illustrated in this case—interstate allocations, water supply agreements, wetland protection, endangered species enhancement, water rights purchase, Newlands Project operation, Fallon Tribe claims settlement—are addressed in the 1990 settlement act. However, many provisions of the bill must still wait for completion of the National Environmental Policy Act process. The preliminary settlement agreement is contingent upon development of an operating agreement for upper Truckee basin reservoirs. A new management plan will also consider instream flows to protect and enhance the resident fishery. Thus the tribe's entitlement to a block of storage water and the ongoing transfer of water rights from agriculture to urban and environmental uses are forcing the Bureau of Reclamation to seek a river basin management solution to the disputes. A dominant factor in the development of OCAP was the Endangered Species Act—final OCAP were required to pass the test of "not likely to jeopardize" the cui-ui and Lahontan cutthroat trout. The Bureau of Reclamation has endorsed and worked diligently to achieve negotiated and legislated solutions to the water disputes in the Truckee-Carson basin. Throughout the consultation process, the Bureau of Reclamation worked with the U.S. Fish and Wildlife Service to develop computer models that could assess the relative impacts of various water management (storage and delivery) plans on cui-ui population survival. This cooperation continues and provides the basis for the yearly OCAP.

As part of its recent shift toward a management emphasis in its mission, the Bureau of Reclamation will need to recognize that there is a disparity between regulations governing water management and historical practices; the agency that controls the water controls other resources as well *and* has the responsibility and authority to protect social and environmental values (in that the federal government owns the storage and distribution facilities but not the water and so cannot market the supply). To function as a water resource management agency, the bureau must view itself as such and truly espouse a "total water management philosophy" among multiple users, where project operating efficiency and local drain-water quality will be primary parameters for establishing basin management criteria and seasonal delivery patterns.

REFERENCES

Athearn, T. G. 1986. The Mythic West in Twentieth-Century America. Lawrence: University of Kansas Press.

Hays, S. P. 1959. P. 11 in Conservation and the Gospel of Efficiency: The Progressive Conservation Movement 1890-1920. Cambridge, Mass.: Harvard University Press.

Hurt, R. D. 1987. Indian Agriculture in America: Prehistory to the Present. Lawrence: University of Kansas Press.

In re Manse Spring and Its Tributaries, 108 P.2d 311, Nevada (1940).

Knack, M., and O. Stewart. 1984. As Long as the River Shall Run: An Ethnohistory of the Pyramid Lake Reservation. Berkeley: University of California Press.

Nevada v. United States, 463 U.S. 110 (1983).

Pyramid Lake Paiute Tribe of Indians v. Morton, 354 F. Supp. 252, D.D.C. (1973).
Pyramid Lake Tribe of Indians v. Hodel, 882 F.2d 364, 9th Cir. (1989).
Reisner, M. 1986. P. 116 in Cadillac Desert. New York: Viking Penguin Press.
State v. Morros, 776 P.2d 262, Nevada (1988).
Truckee River Atlas. 1991. State of California, Department of Water Resources. Sacramento.
U.S. Department of the Interior (U.S. DOI), Bureau of Reclamation. 1986. Pp. 68-69 in Draft Environmental Impact Statement for the Newlands Project Proposed Operating Criteria and Procedures. Sacramento: URS Corporation.
United States v. Alpine Land and Reservoir Co., Nos. 87-1746; 87-1747, U.S.D.C. Nev. (1980, aff'd.).
United States v. Alpine Land and Reservoir Co., 878 F.2d 1217, 9th Cir. (1989).
United States of America v. Orr Water Ditch Company, Final Decree (1944).
Yardas, D. 1987. Birds Versus Fish: An Environmental Perspective on Water Resource Conflicts in the Truckee-Carson River Basins. Comments prepared for the Water in Balance Forum, Reno, Nev.

6

Colorado Front Range-Arkansas River Valley: Interconnected Water Sources

A decision to take water from one source is inherently a decision to leave another source alone—at least for that party at that time. The consequences, both negative and positive, reverberate across political and hydrologic boundaries. The Colorado Front Range-Arkansas River valley illustrates how complex the interconnections can be among water storage and delivery systems and between ground and surface water use. It also illustrates the strengths and weaknesses of robust water markets.

Along the Front Range, water is allocated by a variety of entities that control large quantities of ground and surface water. Many different entities want water in many places, and the transferability of water rights has created a huge statewide market, but no mechanism exists to assess and regulate the market when private and social costs diverge. Political accountability is lacking because the state has traditionally abdicated any role in water planning. Thus, the market operates without regard for consequences except to those interests protected by law: water rights holders.

The effects on third parties—people, communities, and environmental interests—are beyond the scope of considerations legally required of state decisionmakers. This limitation is particularly acute in Colorado because transfers are controlled entirely by the law of prior appropriation; no judge or administrator is authorized to apply public interest criteria. The result is an inefficient system driven by a single allocation strategy: the pressure to acquire more water rights.

Large quantities of water are held by some entities who have earmarked it for future growth needs, while other entities can barely meet present demand with available supplies.

Colorado's allocation ethic is the product of the region's effort to provide stable water supplies in the face of the variable ones provided by nature. In a semiarid area, periodic multiyear droughts create strong pressures to acquire sufficient water rights to assure adequate water supplies during water-short years, as well as for future growth. In anticipation of agricultural growth, federal projects were built to augment available supplies along the Front Range by importing water from other regions. This produced abundant supplies for agriculture. These systems were expanded to provide additional imported water for urban growth. In general, the Front Range has historically followed a pattern of importing more and more water rather than reallocating water from existing agricultural uses to higher-valued municipal and industrial ones.

On the surface, the Colorado Front Range (CFR) cities' approach may seem illogical: they first claim and develop the most distant unappropriated water and only later resort to transfers of nearby developed agricultural rights. Thus, the sources of water for CFR growth historically have not been the easiest and least expensive to develop. The most likely sources of water would seem to be the farming areas close to growing cities, but most of the water sought by cities such as Denver, its suburbs, and Colorado Springs has been from distant watersheds like the Colorado River basin and the Arkansas River valley. The pattern reflects the water rights system's incentives to claim and develop the resource as rapidly and fully as possible.

This phenomenon is explained in part by a desire to avoid a political backlash from dewatering the most productive farming areas in the state. A less obvious explanation is that developing water in the Colorado River watershed—the west slope—gave the cities senior rights in water that has not been developed for a "beneficial use." The Colorado Supreme Court encouraged this by tolerating long lead times to allow cities to design and construct municipal transwatershed projects. The beneficial use doctrine is supposed to curb speculation in water rights by tying the right to use, but it did not perform this function in the CFR. Although transbasin imports are more expensive than the transfer of nearby agricultural water, imports give the cities (and the state generally) more control over large quantities of "unused" water as well as providing a more stable water supply for the region and extending the life of local agricultural economies. In effect, this strategy allowed urban areas to bank future supplies on the farm and then withdraw them as needed. This has been the case in the Arkansas Valley. A similar pattern is appearing in the Denver metropolitan area as west slope sources are more fully claimed and developed. The cities are now pursuing transfers from agriculture.

Colorado presents a classic example of how growing demand for urban water affects rural areas that are a source of "new" water for the cities. Until recently, external effects have not been a significant concern for water developers and users. Even the rural areas where water originates accepted the fact that the state had a strong interest in developing all sources of supply before the water flowed out of state. As undeveloped water supplies dwindle and the west slope becomes more conscious of the effects of the depletion of the headwaters and the recreational and amenity value of these waters, sentiment for redress of third party impacts eclipses the theoretical benefits of keeping the water in state.

In recent transactions and water development proposals, transfer opponents have found ways to force consideration of their interests. Recent experiences with transbasin diversions and transfers of agricultural water have demonstrated a potpourri of ad hoc, but occasionally powerful, levers that can be used by these third parties, such as the application of federal laws, local government powers, and litigation. Unfortunately, these actions do not operate in a consistent and predictable way, and they can impede or prevent transfers or developments that are, on balance, socially and economically beneficial.

Thus, third party effects are now playing an important, although not always decisive, role in whether, where, and how water is developed in Colorado. But third party influence varies with the type and relative wealth of the developer and the federal permits required. There is no state law or system for considering how water resources can be developed most efficiently and with the least impact. Nor is there an integrated mechanism for weighing, avoiding, or mitigating the impacts on third parties. The results are uncertainty for both developers and affected parties, inefficient allocation of some water resources, excessive costs, and avoidable, uncompensated third party effects.

THE SETTING

The term Colorado Front Range refers to a 30- to 40-mi-wide (50- to 60-km-wide) area just east of the Rocky Mountains, extending from Walsenburg in the south to Wellington in the north. The estimated 1989 population of the 10 CFR counties was 2.7 million (approximately 80 percent of the state population). It had been one of the nation's fastest-growing areas, increasing about 2.5 percent per year from 1970 to 1980. The growth rate decreased to 1.7 percent per year in the 1980s (Colorado Department of Local Affairs, 1990), after the energy boom collapsed.

The largest cities in the Front Range are Denver (498,000), Colorado Springs (283,000), and Aurora (188,000). About 65 percent of the state's population is concentrated along the South Platte River after it enters the plains southwest of Denver. The two major river basins in the CFR are the Arkansas in the southeast and the South Platte in the northeast (Figure 6.1). About 70 percent of the streamflow in both basins is from snowmelt, 60 percent of which occurs in May and June.

Arkansas River Basin

The seven counties of the Arkansas River basin had an estimated 1989 population of 210,000, approximately 6 percent of the state (Colorado Department of Local Affairs, 1990). The Arkansas River originates in the central part of the state and flows southward for 75 mi (120 km), then eastward for about 160 mi (260 km) into Kansas. The major water conservancy district in the Arkansas River basin is the Southeastern Colorado Water Conservancy District (SCWCD). It depends on the Frying Pan-Arkansas Project to get water from the Colorado River basin.

Large-scale irrigation began near Rocky Ford in 1874. During a 7-year drought in the 1950s, water transfers, initially to the city of Pueblo, began in the lower Arkansas. The prolonged drought devastated the local economy and caused the closure of the Sugar City sugar factory in 1967. Major sales of water rights have continued to the Pueblo West Metropolitan District, Colorado Springs, Aurora, and Public Service Company (MacDonnell et al., 1990).

South Platte River Basin

The South Platte River basin occupies approximately 18 percent of the state. The South Platte begins in South Park and flows about 270 mi (440 km) northeast to the Cache la Poudre, a principal tributary, then east to Julesburg and into Nebraska. In the mid-1800s the flow in the South Platte disappeared into the river sands from Fort Morgan to the North Platte in Nebraska during July, August, and September. Early explorers reported that there were no trees along the river. By 1910, because of irrigation development and return

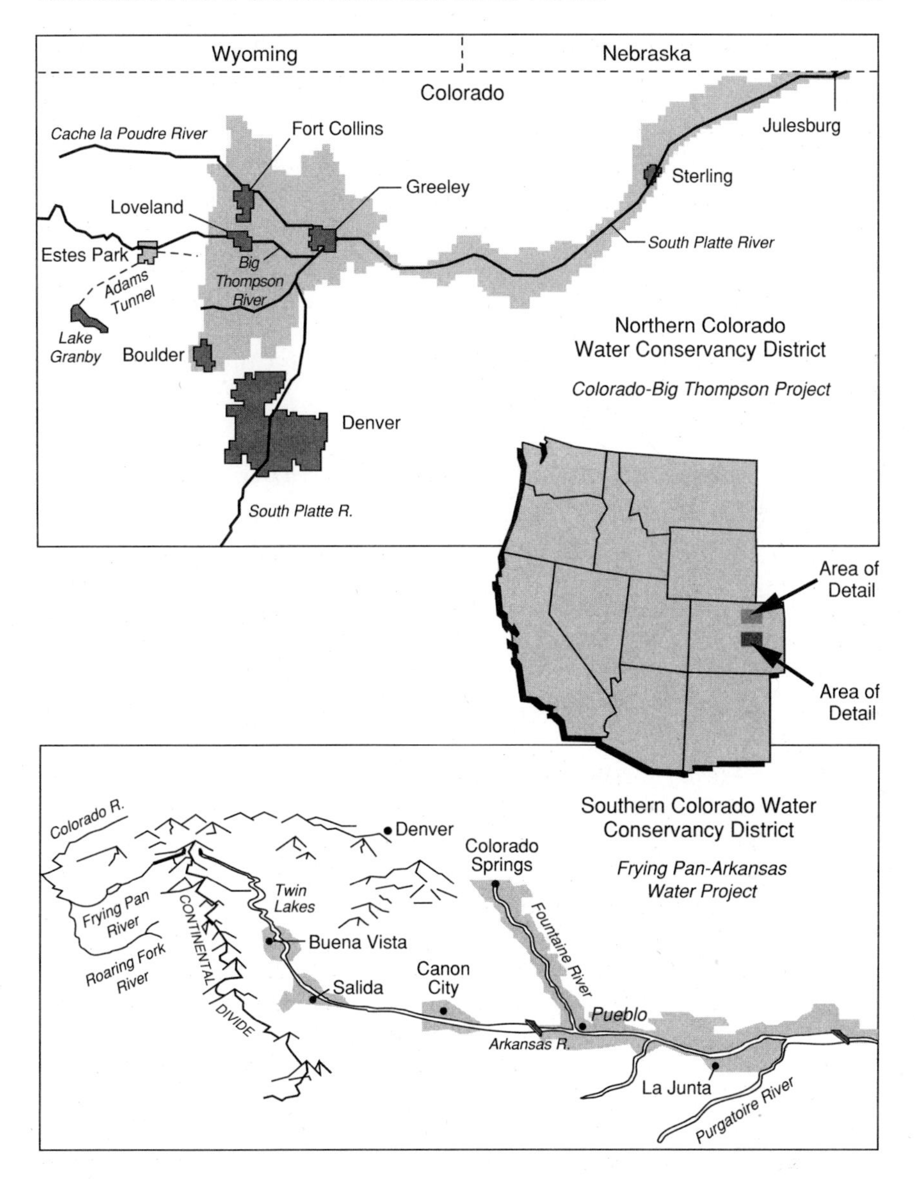

FIGURE 6.1 Water conservancy districts in the Colorado Front Range. SOURCE: U.S. Geological Survey (1986).

flows, the flow in the river continued throughout the year at the state line (GASP, 1989). Since then, with further development of irrigation and importation of water from the Colorado River, the year-round flow has increased.

The Northern Colorado Water Conservancy District (NCWCD) is the major water conservancy district in the South Platte River basin. The district was formed in the 1930s to contract with the Bureau of Reclamation to serve the water needs on the east slope of the continental divide by using the more abundant water supplies from the west slope. The NCWCD depends on the Colorado-Big Thompson Project for interbasin transfers of water from the Colorado River basin.

Areas of Origin for Transmountain Diversions

The major area of origin for transmountain diversions is the upper Colorado River headwaters region on the west slope of the Rocky Mountains. Large quantities of water are imported to the CFR, into both the South Platte and the Arkansas river watersheds, from the Colorado River watershed. Lesser amounts are derived from the North Platte River basin. A proposal exists to transfer fossil ground water to Denver from the San Luis Valley in the Rio Grande River basin.

Early in this century, importing districts and cities began to appropriate rights to Colorado River water for future growth, beginning with the large appropriations by Denver. Extensive diversion, pumping, and transmission facilities have been built in the headwater areas of the tributaries of the Colorado River. They currently convey about 0.5 million acre-feet (0.62 million megaliters (ML)) of water per year from the west slope across the continental divide to the CFR through several tunnels. Reservoir storage has been built on the west slope for the benefit of the CFR.

The waters of the Colorado River are allocated by interstate compacts and U.S. Supreme Court decrees. The degree of use or nonuse by users in individual states cannot affect the state's share. The 1922 Colorado River Compact was designed specifically to ensure that the river would not be legally allocated to faster-growing states such as California to the detriment of Colorado and other upper basin states—then sparsely populated areas dependent on mining, irrigated agriculture, and livestock ranching. The compact guarantees that a share of the water could be used in Colorado if and when needed. In the meantime, the upper basin states must release water as long as lower basin states have beneficial use for it and upper basin states do not. Thus, if additional upper basin storage facilities were built without upper basin beneficial uses, they would be operated essentially for the benefit of the lower basin (Getches, 1986).

East slope-west slope tension has been an enduring factor in political life, but the need for formal area-of-origin protection did not arise until after the Supreme Court adopted the one-person, one-vote standard for legislative representation. Colorado legislation designed to protect areas of origin from water exports was passed in 1973. It applies only to exports by conservancy districts (i.e., not to importing cities such as Denver or Colorado Springs) from the Colorado River basin. In practice, exporting conservancy districts usually build "compensatory storage" projects within the Colorado River basin, sufficient in size to store the volume of water diverted out of the basin annually. This practice is considered by some to be wasteful and of little immediate tangible benefit to the area of origin because compensatory storage projects have stood largely unused for many years (Getches, 1988).

MAJOR WATER TRANSFER PROJECTS

Colorado-Big Thompson Project and Windy Gap Project

Two major transmountain diversion projects bring Colorado River water to the NCWCD—the Colorado-Big Thompson (C-BT) Project and the Windy Gap Project. The C-BT was the first and largest transmountain diversion project in Colorado. It was authorized in 1937, the Bureau of Reclamation began construction in 1938, and the first water was delivered in 1947. The project was completed in the late 1950s. The C-BT was unique in that it was designed to provide supplemental water to an already developed area. It would deliver up to 310,000 acre-feet (382,000 ML) of water annually into the St. Vrain, Little and Big Thompson, and Poudre rivers for agricultural, municipal, and industrial users.

The locally financed and more recently completed Windy Gap Project began deliveries in 1985. It provides rights to an additional 48,000 acre-feet (59,200 ML) of transmountain diversion water for municipal and industrial users in a special subdistrict of NCWCD. A significant portion of the water available from this project is not yet being used, although cities within the district are negotiating to transfer rights from holders of surplus rights from the project. At this point, transfers to cities outside the district are not permitted by district law.

The Hansen canal delivers water to the Cache la Poudre River via the Colorado-Big Thompson project. The Colorado-Big Thompson project was the first and largest transmountain diversion project in the state. CREDIT: Northern Colorado Water Conservancy District.

Arkansas River Basin Projects

More than a dozen transmountain diversion projects have been built to serve the Arkansas River basin, although none approach the size of the C-BT. Most are small projects owned by individual municipal water service organizations. The two major exceptions are the Twin Lakes and Frying Pan-Arkansas (Fry-Ark) projects, built by the Bureau of Reclamation.

The Fry-Ark Project, managed by the SCWCD, is the largest transmountain diversion works in the Arkansas River basin. It delivers approximately 80,000 acre-feet (99,000 ML) of water to agricultural, municipal, and industrial users in the SCWCD. About 70 percent of the water comes from transmountain diversions of Colorado River water, and the remaining 30 percent is developed from storable floodwaters of the Arkansas River and its tributaries (SCWCD, 1981).

The Twin Lakes Project is owned in part by the federal government and in part by the Twin Lakes Reservoir and Colorado Canal Company. The shares of mutual stock that may be bought and sold on the market represent only that portion of the water in the reservoir controlled by the private water company. All remaining storage space in the reservoir is used to support the Fry-Ark Project. Roughly 10 percent of the total average annual yield of Twin Lakes stock is attributed to native flow, and the rest represents transmountain diversion water rights deeded to the company by the Bureau of Reclamation (Saliba and Bush, 1987). About 50,000 acre-feet (62,000 ML) of water per year is delivered to Twin Lakes company stockholders, primarily the cities of Aurora, Colorado Springs, Pueblo, and Pueblo West (Saliba and Bush, 1987).

Infrastructural Differences Between NCWCD and SCWCD Projects

The water service infrastructure within the NCWCD is more integrated and extensive than that within the SCWCD. The system of capturing, storing, and distributing water resources in the NCWCD is one of the most complex water supply networks in the western United States. Dozens of different water service organizations operate hundreds of miles of canals, storage reservoirs, pumping facilities, and water treatment plants serving an area roughly the size of Connecticut. In the Arkansas River basin, water storage and delivery facilities are less sophisticated, and the opportunities for exchanging water among the many different and autonomous organizations are correspondingly more limited.

The Fry-Ark and C-BT were both conceived as multipurpose projects to supply supplemental water to existing irrigated lands and growing urban centers. The characteristics of each project are quite different, however. Shortly after its creation in the late 1950s, the governing board of directors of the SCWCD decided that the allocation of Fry-Ark water would be fixed at 51 percent for municipal and industrial use and 49 percent for irrigation. The allocation may be changed by the board in the future to reflect conversion of agricultural lands to nonagricultural use. Users of Fry-Ark water who sell other (non-project) water rights that they hold are not permitted to replace those rights with Fry-Ark water (SCWCD, 1981). The unit price for Fry-Ark water is predetermined by the board and is the same for all users. The SCWCD holds both the primary and the return flow rights to Fry-Ark water, and individual users may not resell their project water (Saliba and Bush, 1987).

Fry-Ark water users may store their Fry-Ark entitlements in project reservoirs. Agricultural users may hold their unused water until May 1 of the following year, and municipal and industrial users may carry their unused water over from year to year (Saliba and Bush, 1987). The SCWCD also allows agricultural users to store their own (non-project) decreed native flow rights during the winter season for use in the summer (SCWCD, 1985).

Contractual shares in the water rights under the C-BT are held directly by the individuals and water service organizations within the project service area. The NCWCD retains all rights to the return flows of project water, but each water user has the full right to purchase, sell, trade, or rent rights to the primary flows (U.S. Bureau of Reclamation, 1938). Return flows from the C-BT may be neither recaptured nor resold by C-BT users. Return flows that are not allocated by the NCWCD to its users simply remain in the river for others downstream. The effect has been to firm up the water supply available under native flow appropriations on the lower reaches of the South Platte because the imported water becomes just like other return flows once it is returned to the stream. As a result, the flow in the South Platte has increased substantially since 1950. Water users at the downstream end of the NCWCD have found little advantage in holding shares in a project that effectively provides them with water whether they participate directly in it or not. Consequently, many downstream appropriators have sold most, if not all, of their C-BT rights to upstream users (Harrison, 1984).

INSTITUTIONAL AND LEGAL CONSIDERATIONS

In its 1876 constitution, Colorado adopted the prior appropriation doctrine. Changes in use, which accompany most transfers, are reviewed in water court to determine if there is injury to these vested rights. Third parties can participate in these proceedings; however, appropriations and transfers ordinarily are approved unless there is harm to water rights.

Colorado has a long history of water transfers. About 85 transfer applications seeking a change of use were processed each year from 1975 to 1984 (MacDonnell et al., 1990). In normal years, agricultural water users in many conservancy districts are able to rent or purchase water from others within the same district as needed to mature a high-value crop if their own water supplies are limited. In some areas, such as the NCWCD, individuals entitled to use water in excess of crop needs in one year can rent water to someone else within the district who must simply complete and mail a postcard. In other areas, approval of a "temporary substitute supply plan" by the state engineer is required for temporary transfer of water.

Ground water hydraulically connected with surface water streams is administered in the same general way as surface water in Colorado. Deep bedrock ground water that is not hydraulically connected to surface streams is administered by the state engineer, but outside of the prior appropriation system. Ground water in certain "designated basins" is administered by the Colorado Groundwater Commission under a special statutory system.

Neither the state engineer nor the water court historically has dealt with water quality issues as they relate to the use of water rights, although these issues are arising more frequently. The Colorado legislature, however, has passed several laws subordinating water quality regulations to the diversion of water; that is, diversion of water for beneficial use cannot be restricted for water quality protection.

Colorado generally has not planned for the development or management of its water resources. Instead, it has chosen to develop highly technical acquisition rules administered by the courts. Neither policy direction nor consistent technical data are provided as guidance to the state engineer, water court, or water users. State direction is lacking on the amounts of water needed for various present or projected uses by unit area or on the quantity per unit output. Data bases on water rights, water diversions, and control structures along the South Platte River are being developed, however, for use as decision-support systems by the state engineer and others.

CURRENT WATER TRANSFERS AND WATER MARKETING

Water marketing in northeastern and southeastern Colorado has differed in several important ways. In northeastern Colorado, buyers and sellers generally have been within the NCWCD service area as required by district law, whereas market participants in southeastern Colorado have not been exclusively within the SCWCD area. Users within NCWCD can transfer large quantities of water rights over the wide geographical areas included in the district, though not outside the district, by signing a water stock certificate. In southeastern Colorado, however, most transfers require formal water court proceedings to change the point of diversion and place of use of the water rights. Appropriative water rights represented by water company stocks having a large service area, and approval for multiple uses can be transferred readily within their service area at low costs. As is apparent from other types of transactions in the NCWCD and SCWCD, market transfers can also be subject to expensive, time-consuming, and complex approval procedures and litigation.

Northeastern Colorado

The single largest source of water in northeastern Colorado, the C-BT, is also the easiest type of water to transfer. Municipal water service organizations and rural-domestic water companies within the district have generally been able to acquire C-BT units from irrigators without applying to water court to change points of diversion, places of use, or purposes of use.

When units of C-BT water were distributed to project participants in the 1950s, nearly 85 percent was assigned to agriculture. The 15 percent assigned to other uses was adequate to meet virtually all nonagricultural demands, and there was little pressure to reallocate supplies. The market price for C-BT units became established in the early 1960s at a price of about $125 per acre-foot ($100 per ML) (Howe et al., 1986). Through the 1960s and 1970s, prices for C-BT water rights rose. Prices reached $3,600 per acre-foot ($2,920 per ML) by 1980. After 1980, prices began to decline rapidly (Figure 6.2). The price decline had various contributing causes: the completion of the Windy Gap Project brought significant new supplies to the CFR, many people began to believe that past acquisitions were adequate to support growth, and people began to worry about drying up too much local farmland.

The Windy Gap Project was planned and funded locally through a subdistrict formed by the cities of Boulder, Estes Park, Fort Collins, Greeley, Longmont, and Loveland. Participants in the Windy Gap Project share the responsibility for project costs and, in return, share the water supply. Owners of Windy Gap water can use their primary flow rights and then reuse or sell their return flow. The cities have little use for the water and therefore have sold some of their

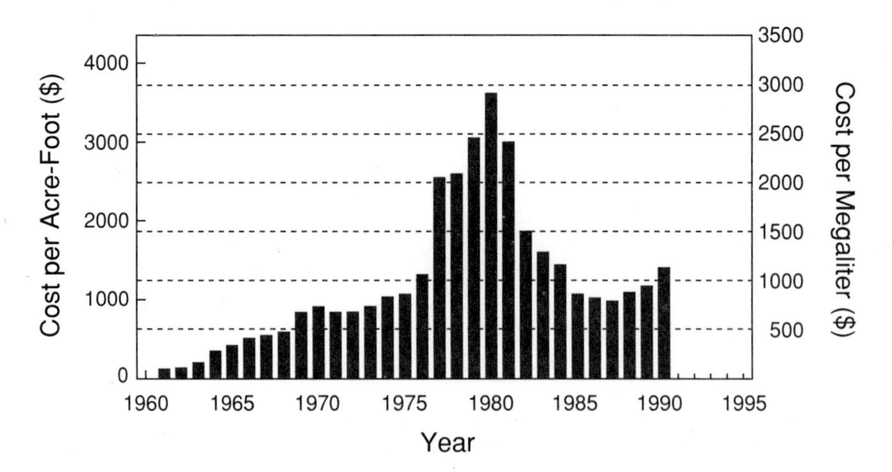

FIGURE 6.2 Water rights prices (1987 constant dollars), Colorado-Big Thompson Project. SOURCE: Colby (1991).

rights, though they are restricted to in-district sales. Since the project's inception, Fort Collins has transferred its direct interest in the project to the Platte River Power Authority, and Estes Park has sold part of its interest to the city of Broomfield and to the Central Weld County Water District.

Proponents of the Windy Gap Project encountered several third party issues that had to be resolved. Because of the area-of-origin law that applies to projects developed by conservancy districts seeking to export water from the basin of the Colorado River, NCWCD was required to build a $10 million "compensatory storage reservoir" to make up for the water that would be exported. That obligation was later converted into a cash payment of $10.2 million.

In the course of negotiations, NCWCD resolved several other issues raised by third parties. Many of the issues had surfaced in the course of preparing an environmental impact statement that was required because the Windy Gap Project would use conveyance facilities of the C-BT Project, a federal project. The final settlement agreement included the following commitments in addition to the $10 million reservoir for the west slope:

- provide 3,000 acre-feet (3,700 ML) for future development in Grand County;
- pay $25,000 to Grand County to study salinity impacts;
- provide 3,000 acre-feet (3,700 ML) to the Middle Park Water Conservancy District;
- pay $420,000 to the town of Hot Sulphur Springs for upgrading its sewage treatment plant because Windy Gap depletions from the river would reduce dilutive capacity;
- donate $550,000 to a fund for the U.S. Fish and Wildlife Service to study the Colorado River squawfish and other endangered fish;
- undertake certain measures to protect wetlands as suggested by the U.S. Environmental Protection Agency;
- bypass minimum flows of water needed for fish as requested by the state Division of Wildlife;
- subordinate district rights in the Fraser River to those of local agricultural users; and
- pay more than $500,000 for downstream ranchers' headgates that would be affected by the project.

Windy Gap water rights are among the most attractive, marketable water resources in Colorado, but an active market is hindered by the high cost of participation in the project and the district's restriction on transfers to purchasers outside its boundaries, such as the Denver suburbs, where the demand for water is greatest. The current annual cost of water to project participants, including bond debt service and operation and maintenance costs, is between $200 and $300 per acre-foot ($247 and $370 per ML) (Harrison, 1984). A transfer also usually requires construction of delivery facilities.

Southeastern Colorado

The southeastern Colorado water market is at once much simpler and more difficult to describe than the market in northeastern Colorado—simpler because there has been much less market activity than in the NCWCD and more difficult because transfers have involved extensive negotiation and litigation. Transmountain diversion projects bring a total of 200,000 acre-feet (247,000 ML) of water per year into the Arkansas River basin, but only the 50,000 acre-feet (61,700 ML) of this water provided through the Twin Lakes Reservoir and Colorado Canal Company has been marketed.

Twin Lakes Reservoir and Colorado Canal Company water is marketed by selling shares of mutual water stock; rights are divided into native flow and transmountain diversion water rights. The approximately 45,000 shares of stock in the Twin Lakes Company were all originally owned by farmers within the service area of the Colorado Canal Company, located about 50 mi (80 km) downstream of the city of Pueblo along the Arkansas River (Saliba and Bush, 1987). Until about 1970, farmland in the service area sold for about $500 per acre ($1,235 per hectare), including land and all water rights. In the early 1970s an investment group offered farmers in the area about $900 per acre ($2,225 per hectare). Despite heated local opposition, many farmers in the area sold their land and water company stock. As a result, slightly more than 60 percent of the Twin Lakes stock changed hands. The remaining stockholders formed a coalition that became known as the Proxy Group. The investors and the Proxy Group successfully obtained a decree in water court to change the purpose of use for Twin Lakes water rights from irrigation to multiple use (Colorado District Court, 1974). Twin Lakes stock suddenly became one of the most flexible and valuable sources of water in the area, and nonagricultural users quickly bid up its price.

Between 1972 and 1975, nearly all shares of Twin Lakes water stock were sold to the cities of Pueblo, Pueblo West, and Colorado Springs for prices ranging from about $2,300 to $2,400 per acre-foot ($1,865 to $1,945 per ML). Small quantities of stock have reportedly been sold to homeowners in mountain resort communities for prices ranging from $8,000 to more than $10,000 per acre-foot ($6,500 to more than $8,000 per ML) (Saliba and Bush, 1987).

In the fall of 1986, Colorado Springs purchased 17,500 acre-feet (21,600 ML) of direct flow and storage rights in a complex transaction entitling the city to store unused water from some of its direct flow water rights in the Henry and Meredith reservoirs should there ever be insufficient storage capacity in Pueblo Reservoir. Water stored in these reservoirs is then available to serve downstream users, who in turn may exchange their flow rights to Colorado Springs for water upstream in the Pueblo Reservoir.

The city of Pueblo, located along the Arkansas River just below Pueblo Reservoir, purchased shares of stock in other ditch companies and has built some water development projects of its own. In recent years, Pueblo's principal water company acquisitions have been the purchase of storage rights in the Otero Canal Company and the Booth-Orchard Canal Company and the purchase of both storage and flow rights in the Rocky Ford Highline Canal Company (Saliba and Bush, 1987).

The city of Aurora, a suburb east of Denver that experienced rapid growth until the mid-1980s, concluded three acquisitions of water rights from the basin at the close of 1986. A majority interest in the Rocky Ford Ditch Company was acquired, giving the city 8,200 acre-feet (10,000 ML) of water at a cost of about $2,500 per acre-foot ($2,000 per ML). Aurora also bought 5,600 acre-feet (6,900 ML) from the Colorado Canal Company, at a cost of approximately $2,500 per acre-foot ($2,000 per ML). Finally, Aurora acquired 45 percent of the outstanding shares of stock in the Busk-Ivanhoe Ditch Company, yielding about 3,000 acre-feet (3,700 ML) at a cost of $3,500 per acre-foot ($2,840 per ML). The city was willing to pay a premium price for the Busk-Ivanhoe stock because it included transmountain diversion rights, which are both legally and hydrologically easier to transfer to the South Platte River basin than are native flow and storage rights in the Arkansas River, and which can be consumed (Saliba and Bush, 1987).

FUTURE TRANSBASIN DIVERSIONS

Numerous reservoirs and an extensive water distribution network allow great flexibility in using and managing transferred water. For the foreseeable future, no significant increase in agricultural demand for water in the CFR is predicted. The primary demand for increasing or reallocating water supplies will be municipal growth. Demographers continue to predict that population, and thus demand for water, will increase in CFR urban areas. Although projections are more modest than in the past, and actual growth has decreased because of a 10-year economic slump, future expansion is still expected. Continued municipal and industrial growth in water demand will maintain some pressure to transfer existing agricultural water rights. Major proposed or pending transfers are (1) within the NCWCD, (2) from the irrigation districts in the Arkansas River basin to Colorado Springs, (3) from the NCWCD to the city of Thornton, (4) from the San Luis Valley in south central Colorado to municipal use in the CFR, and (5) from Colorado River tributaries such as the Gunnison River to the CFR.

Conditional water rights already exist for several transbasin diversion projects to bring water to the CFR. One of them is a large diversion known as Homestake II that would take water from Eagle County to Aurora and Colorado Springs. Significant west slope opposition to the project based on environmental, land use, and economic impacts has led to major legal challenges. Although opponents lost in water court, they continued their opposition in other forums. Using county powers under an otherwise weak state land use law, Eagle County halted construction of the project for the time being.

Recent proposals of the city of Aurora and Arapahoe County would remove large quantities of water from the Gunnison River. Local residents have joined with environmentalists throughout the region to oppose dewatering of the river. Faced with this strong opposition, Aurora dropped its proposal. The water court is considering the Arapahoe County case, in which the opponents are raising both traditional water rights issues and effects on third party interests. If the court considers matters such as fish and wildlife impacts, recreational effects, consequences for the local economy, detriment to public uses, and other public value or public trust arguments, it will be the first time in Colorado water jurisprudence.

The largest project proposed to develop water for the CFR is Two Forks, a 1.1-million-acre-foot (1.36-million-ML) reservoir on the South Platte River that would flood about 30 mi (50 km) of the river and its North Fork. About two-thirds of the water to supply the project's 98,000-acre-foot (120,000-ML) yield would originate from the Colorado River basin. The project was opposed vigorously by west slope interests, sportsmen, area residents, and environmentalists. During a lengthy and expensive ($40 million) environmental impact statement process, required by federal law because the dam needed a federal permit, an elaborate mitigation plan was developed. The plan, with some modifications, was incorporated as conditions in the permit approved by the U.S. Army Corps of Engineers.

The Two Forks mitigation plan was a highly detailed description of recreational facilities, trails, picnic and campgrounds, boat ramps, highway relocations and improvements, and relocation aid for residents. It provided for improvements to fishing streams, fish-stocking programs, flow modifications, riverbed and fishery enhancement on the west slope, big game habitat improvements, wetlands rehabilitation, protection of archeological sites, water quality monitoring of affected streams in both watersheds, and attention to endangered species problems. A variety of adverse residual environmental and social effects were identified in the environmental impact statement.

There was no state process to deal with the far-reaching impacts of the Two Forks Project, nor any occasion for the water court to review the matter. Denver had long held conditional water rights for the project, and under existing water law water courts consider only effects on other water rights. The only reason there was a forum to voice diverse affected interests was that Section 404 of the Clean Water Act requires a permit to place fill in "waters of the United States." The U.S. Army Corps of Engineers made an extensive review of third party effects and then granted a permit.

The Two Forks permit was vetoed by the U.S. Environmental Protection Agency administrator in late 1990 on the basis of the agency's finding that there were less environmentally damaging alternatives. Like the U.S. Army Corps of Engineers, it tried to integrate state and federal concerns and to reach an appropriate policy choice, but it came to the opposite conclusion. Throughout the process, federal officials expressed frustration over the lack of any state water plan or policy by which state interests could be judged. The governor, when asked for a position on the project, voiced many grave concerns but said that he believed that a long-term permit could give proponents time to pursue alternatives, such as water conservation and cooperative use of existing substantial water supplies. He was uneasy about the destruction of natural resources on both sides of the mountains but said that he had no legal authority to impede or modify the plans of water developers.

A recent major water development proposal would pump large quantities of deep ground water from the San Luis Valley in the watershed of the Rio Grande River and convey it to unspecified customers on the CFR. There is considerable local opposition in the San Luis Valley to the transfer because of the possible effects on existing water users and related socioeconomic impacts. The proposal is being reviewed in water court, where, as has already been noted, objections are confined to allegations of interference with water rights.

THIRD PARTY IMPACTS

Socioeconomic Impacts

AGRICULTURAL SECTOR

Agriculture has been declining in the Arkansas River basin since the 1960s as water has been transferred to cities. Many of the remaining farms and fields are small; the primary crops are melons and vegetables. Continued decline is expected because these are labor-intensive operations, and many younger people do not wish to remain in farming.

Still, the primary economic base of the seven-county area in southeastern Colorado is agriculture and agriculture-related enterprises. Although the transfers from agriculture to municipal use involve willing sellers and buyers, and the direct economic effects are favorable, indirect local and regional economic impacts can be significant; these include (1) reduction in demand for agricultural production inputs; (2) reduction in crop outputs, which provide inputs to other production processes; and (3) reduction in consumption due to reduced income, or multiplier effects. Models of these effects with and without agricultural buyouts generally show that lands removed from irrigation would not be suitable for other dryland agriculture (Howe et al., 1990).

RURAL COMMUNITIES

Rural communities in the area of origin have suffered most from water transfers. In many cases, the communities were in decline prior to the transfers, but transfers accelerated the process. In the Arkansas River basin, the population of the six rural counties of southeastern Colorado in 1930 was 68,576 and the population of Pueblo County was 66,038. The estimated 1987 population was 50,000 for the rural areas and 132,000 for Pueblo County (Howe et al., 1990). This represents a net growth rate of –0.55 percent per year for the rural areas and 1.22 percent per year for the urban area.

A study of Crowley County, where major shares of water were transferred from agricultural lands to Colorado Springs, Aurora, Pueblo, and Pueblo West, illustrates the potential impacts on rural communities (Weber, 1989). There has been little reinvestment of the water transfer proceeds into the local area. Rather, 60 to 75 percent of the sales proceeds went to debt payment and taxes. For many farmers on the verge of bankruptcy, the opportunity to sell water shares provided a better option than possible future foreclosure. In addition, Pueblo and Pueblo West acquired more water than is currently needed and have leased water back to the users in the county. But in time, Crowley County's businesses, social institutions, and nonselling families are expected to pay for the water transfers. Agriculture, which has been declining, was struggling to support a local service sector before the transfers. With the transfers, however, land values have dropped drastically, and the local economy may fall below the threshold of viability.

MINORITIES

The impact of water transfers on minority populations arose under a recently approved change in use from irrigation to mining in the San Luis Valley. The change was bitterly opposed by some members of nearby Hispanic communities concerned about the impacts of mining on the San Luis watershed—site of the oldest water rights in Colorado and of the San Luis Commons, one of the last remaining such lands in the Southwest, a legacy of the early grazing commons of the Hispanic settlers. The change in use was approved, with some conditions to prevent impairment of other water rights. The court allowed community members to express their concerns at a hearing, but Colorado water law does not provide a basis for considering these types of concerns in reviewing a change in water use.

Environmental Impacts

INSTREAM VALUES

Fish and wildlife habitat and riparian vegetation along the Arkansas and South Platte rivers have deteriorated during the past century. Because of controlled, greatly altered flow regimes, the landscape along the rivers has changed dramatically. Old cottonwood groves that developed because of irrigation development and controlled flows provided habitat for raptors, other birds, and a variety of wildlife. Without water, these groves have died out, bringing a decrease in wildlife. Wetlands historically sustained by irrigation, which supported waterfowl, also have dried up. There has been some replacement of lost wetlands and other riparian habitat, however, especially for waterfowl, as old gravel mines along the rivers have filled in with water.

Water transfers within the watersheds of the Arkansas and South Platte rivers have tended to consume water high on the river because urban uses are upstream of the established agricultural areas. This situation reduces the quantity of water available for fisheries, riparian vegetation, and recreational uses downstream. Because the water in the streams is heavily appropriated, releases from reservoirs and water uses are timed to meet established water rights involving large consumptive uses. There is, therefore, little consideration for instream flows except to the extent that instream flow rights are held by the state.

Colorado has an instream flow program that allows the Colorado Water Conservation Board (CWCB) to appropriate rights to sufficient water "to preserve the natural environment to a reasonable degree." The program provides only limited protection for instream flows in the Arkansas and South Platte for a couple of reasons.

First, the rights appropriated by the CWCB are usually quite junior. The instream flow law was passed in 1973. Water in most streams was already appropriated by that date, so the instream water right ensures flows only in years when there is enough water available after all senior rights have been satisfied. The law does provide some protection against changes in point of diversion, however, because the CWCB is treated as any other water user: its rights may not be harmed by a transfer upstream, even if the transfer is of a senior right.

Second, the CWCB has interpreted its authority narrowly, providing minimal instream appropriations to maintain cold-water fish such as trout. Water rights needed to maintain riparian vegetation, wilderness characteristics, and other types of fish and wildlife have not yet been the subject of CWCB appropriations, even though preservation of these values appears to be within the purpose of preserving "the natural environment to a reasonable degree." Because of the CWCB's present narrow reading of the statute, only small appropriations have been made. Another problem is that because fish habitat, and the natural environment generally, were seriously degraded long ago on many parts of these rivers, it can be argued that there is not much to "preserve" at this late date. Thus CWCB instream flows have been appropriated largely on tributaries to the two river systems in the mountains, not in the main stems.

There is growing interest in protecting instream flows in the rivers as they flow through cities. In Denver, for instance, citizens have organized the Platte River Greenway Foundation and developed an attractive system of trails and naturally landscaped park areas along the South Platte River. The Greenway Foundation has built boating facilities and plans to reintroduce fish to parts of the river. Efforts are being made to maintain sufficient flows to give these public amenities their fullest value. These efforts were frustrated at first by the fact that only CWCB can hold instream flow water rights, but difficulties were overcome with agreements regarding upstream dam releases and operations.

CONCLUSIONS

Water Rights Transfers

1. Water rights transfers are commonplace in Colorado. A well-established process for transfers exists in Colorado through the water court. This process has successfully protected those who hold water rights from injury resulting from transfers. The judicial process, however, can be complicated, time-consuming, and expensive. In spite of the time and expense involved, the water court process has not protected all third parties adequately. There is a need to improve the efficiency of the process and make it more accessible.

In the Northern Colorado Water Conservancy District the newer nonagricultural water users are intermingled with older agricultural users, and both are part of the district, so water transfers often are simplified. In contrast to the relatively smooth transfers of water rights within the NCWCD, transfers in the Arkansas River basin often are complex and difficult. Population growth is clustered around Pueblo, Colorado Springs, and Fountain, which are upstream from most existing agricultural water users. Therefore, transferring water rights from irrigators to urban users involves complicated exchanges over long distances and across jurisdictional boundaries. Furthermore, numerous ditch companies have contractual rights to water from the Southern Colorado Water Conservancy District, as well as their own appropriative rights to river water. These companies as well as the district must be reckoned with in any transaction. Although NCWCD transfers to expanding municipal uses within the district can be met with supplies of imported Colorado River water, future transfers in the Arkansas Valley are likely to require retirement of agricultural lands.

2. Water rights exist for instream use. Colorado state water law includes an instream flow protection process. Maintaining instream flow to preserve the natural environment is considered a beneficial use, and water rights for instream flows are held by the Colorado Water Conservation Board and integrated into the prior appropriation system. However, the board has limited the quantity of water on which it will seek rights to the minimum needed for cold-water fisheries. Thus interests such as wildlife, water quality, recreation, wilderness, and aesthetics have no protection beyond that required by fish. Further, the rights held by the board for instream flow are junior to most consumptive use rights of agriculture or municipalities, so in reality they provide protection only against transfers, exchanges, or other water rights changes that would injure the instream flow rights.

3. Water quality effects generally are not considered. Water development interests have resisted consideration of effects on water quality in the process of administration or the transfer of rights.

4. Rural interests are represented only by water rights holders and conservancy districts. Rural agricultural water rights holders in Colorado are given notice of transfers and their water rights are protected in the transfer process. The water court system allows anyone with potential water rights injury to become a party. Legal and engineering costs sometimes bar individual farmers from effective participation in those processes. However, to the extent that individuals' rights are represented by conservancy districts, which usually become parties in water rights transfers, the districts effectively protect these water rights holders. Broader community concerns generally are not addressed by organizations representing individual right holders or by conservancy districts. Some evidence is emerging, however, that indicates a broadening of conservancy district concerns to encompass wider community concerns.

5. The third party effects of water transfers need additional consideration in Colorado. Colorado is the only western state that has no law directing that the public interest be taken into account in some form during water transfers. Expanding consideration of interests beyond the traditional test of no injury to other water rights would be controversial in Colorado and would be strongly opposed by water developers. One way to proceed would be integrate the analysis of third party effects into federal permit requirements (e.g., Section 404 of the Clean Water Act or Section 9 of the Endangered Species Act). This might help avoid duplication and reduce political opposition. At a minimum, acceptable procedures need to be developed to compensate basin-of-origin residents for major transfers of water to urban areas.

6. Basin-of-origin issues are handled on a case-by-case basis. Basin-of-origin issues are resolved on a case-by-case basis involving (1) negotiation outside the water court process, (2) political activities, and (3) the usual water rights appropriation and transfer process. This ad hoc resolution process has produced inconsistent results, but it has sometimes provided compensation or mitigation for the area of origin.

Although there is no mechanism for comprehensively and consistently considering third party effects of water use and development in Colorado, these matters can be raised and resolved in a variety of ways. The level of success depends on the laws that apply, the political strength and sophistication of the parties, and the financial capacity of the project proponent to compensate or make adjustments in the project plans in response to third party complaints.

Several examples in this case study show how case-by-case negotiation works to resolve third party effects associated with transmountain diversions in Colorado. They include notably the Windy Gap and Two Forks projects. Both resulted in major concessions to affected third parties in the area of origin and a redress of local and statewide concerns.

The principal problem with the case-by-case approach is uncertainty. Project proponents do not know the potential costs or modifications required, and affected parties do not know which of their claims will be resolved satisfactorily and which will not. There can be wide variations from one case to the next in how third party effects are resolved. Also, it appears that in the case-by-case approach there is little leverage for third parties outside the context of major project construction; transfers in the absence of construction provide few legal levers for third parties.

Water Resource Planning and Management

1. Basinwide water resource management could be consolidated. Basinwide water resource planning and management could consolidate overlapping authorities and reduce the conflicts and costs involved in water transfers. Integrated water supply planning and management in the CFR area could increase efficiency and cost savings. Perhaps most importantly, integrated water supply planning and management could result in increased water supply as a result of the more efficient operation of combined systems. Unless leading water supply agencies or the state takes the lead in promoting integrated planning and management for water supply in the CFR area, the present "every government for itself" situation will continue.

2. Planning for drought could be enhanced. Colorado law permits temporary and permanent changes in water rights. Contingency plans could be negotiated to provide that farmers would decrease the amount of low-value crops planted and sell available water to users willing to pay higher prices. (This practice is being proposed in the Northern Colorado Water Conservancy District.) Resistance to drought also would be enhanced if policies encouraged reduced water demand. Municipalities could impose water use constraints and increase the cost per unit volume in order to reduce water use without reducing operating revenue. Policies for developing and managing ground water supplies could help stabilize water supplies and provide supplies during drought. Conjunctive use of surface and ground water is practiced, but policies and practices need further development.

3. Water salvage could be encouraged. Colorado law is clear that water being consumed by evaporation or phreatophytes cannot be salvaged by reducing consumptive use and then be sold for use elsewhere. Colorado law is unclear, however, on the right of a farmer to reduce consumptive loss by more efficient agricultural practices or a change in crop and to sell the salvaged consumptive use elsewhere. Eliminating the legal uncertainties and reducing the legal costs associated with water salvage will enhance the potential of water salvage in water management. Critical examination of the technical merits of alternative salvage processes and their potential salvage amounts and costs is an important first step.

4. More efficient processes for water transfers are needed. Increased efficiency in the water transfer process would reduce costs and promote desirable transfers. Development of standard procedures for estimating consumptive use, transit losses, ground water return flow delay times, and similar engineering calculations could avoid their repetition for every transfer and thereby lower costs. The Colorado state engineer's office could promulgate standardized procedures and coefficients for water transfers.

5. The Colorado-Big Thompson Project could serve as a model for transfers of federal project water. The transfer of water within the NCWCD is a model for water transfers involving federal project water. The existence of the market for Colorado-Big Thompson Project shares and the ease of both short-term and long-term transfers of water provide an example of an efficient federal project water transfer system. The district could lead the way in removing barriers to out-of-district transfers as well. However, as with many other local water districts around the West, proposed transfers of district water to locations outside the district have been opposed by the NCWCD. These barriers to exporting water from a district can prevent transfers from providing a least-cost water supply for other users in the region.

REFERENCES

Colby, B. G. 1991. Water rights price trends: Prospects for the 1990s. University of Arizona, Tucson, Appraisal Journal.

Colorado Department of Local Affairs. 1990. Colorado Municipal Population Estimates, 1981-1989. Denver, Colo.

Colorado District Court. 1974. Water Division No. 2, Case No. W-3965. Twin Lakes Reservoir and Canal Company.

Getches, D. H. 1986. Meeting Colorado's water requirements: An overview of the issues. Pp. 1-24 in L. J. MacDonnell, ed., Tradition, Innovation and Conflict: Perspectives on Colorado Water Law. Boulder: University of Colorado, Natural Resources Law Center.

Getches, D. H. 1988. Pressures for change in western water policy. Pp. 143-164 in D. H. Getches, ed., Water and the American West. Boulder: University of Colorado, Natural Resources Law Center.

Groundwater Appropriators of the South Platte River Basin (GASP). 1989. Pp. 1-31 in South Platte River Compact. Fort Morgan, Colorado.

Harrison, C. 1984. Colorado water marketing: The experience of a water broker. Presented at the annual meeting of the Western Agricultural Economics Association, San Diego, Calif.

Howe, C. W., D. Schurmeier, and W. Shaw. 1986. Innovations in water management: An expost analysis of the Colorado-Big Thompson Project and the NCWCD. Unpublished manuscript. Boulder: University of Colorado, Department of Economics.

Howe, C. W., J. K. Lazo, and K. R. Weber. 1990. The economic impacts of agriculture-to-urban water transfers on the area of origin: A case study of the Arkansas River valley in Colorado. American Journal of Agricultural Economics 72(5):2300-2304.

MacDonnell, L. J., C. W. Howe, and T. A. Rice. 1990. Transfers of water use in Colorado. Pp. 1-52 in The Water Transfer Process as a Management Option for Meeting Changing Water Demands. Vol. II. Award No. 14-08-0001-G1538. Reston, Va.: U.S. Geological Survey.

Saliba, B. C., and D. B. Bush. 1987. Water Marketing in Theory and Practice: Market Transfers, Water Values and Public Policies. Studies in Water Policy and Management, No. 12. Boulder: Westview Press.

Southeastern Colorado Water Conservancy District (SCWCD). 1981. Water Allocation Policy, As Amended. Pueblo. October 22.

Southeastern Colorado Water Conservancy District (SCWCD). 1985. Operating Plan: Winter Plan 1985-1986. Pueblo. November.

U.S. Bureau of Reclamation. 1938. Repayment Contract Between the Bureau of Reclamation and the Northern Colorado Water Conservancy District. Loveland, Colo.

U.S. Geological Survey. 1986. National Water Summary 1985—Hydrologic Events and Surface Water Resources. Water Supply Paper 2000, 516 pp.

Weber, K. R. 1989. What Becomes of Farmers Who Sell Their Irrigation Water? The Case of Water Sales in Crowley County, Colorado. Grant No. 885-054A Report. Ford Foundation. November 16.

7

Northern New Mexico: Differing Notions of Water, Property, and Community

Three distinct cultures—Indian, Hispanic, and Anglo—compete for the scarce water resources of the spectacular but often harsh landscape of northern New Mexico. Each culture has a distinct water allocation tradition. In the nineteenth century, Anglo property concepts were superimposed over the more communal traditions of the pueblos and Hispanic irrigation communities. Today New Mexico has a sophisticated water allocation system that basically treats water as a commodity to maximize the efficiency of use of the resource. But the clash of cultures makes northern New Mexico special; there are allocation tensions that do not exist in other states. Although the Anglo allocation system has dominated for more than a century, communal traditions have survived in the face of this superior economic and political power and now show signs of resurgence. Hispanic villages and Indian Pueblos of great antiquity have survived after New Mexico passed from Spain to Mexico to the United States and are now aggressively trying to preserve their pre-Anglo culture.

The culture's underlying belief in community values, the respect for land and nature, and the unity of life, spirit, art, and work have a special appeal. Ironically, modern efforts by Hispanics and Indians to chart their destiny built on the efforts of earlier Anglo romantics. Post-World War I progressives such as D. H. Lawrence, Mabel Dodge Luhan, Mary Austin, Edgar Lee Hewett, and John Collier found the communal rural life a redemptive alternative to urban life (Forrest, 1989), and another wave of interest among artists, spiritualists, and others seeking its lifestyle is occurring today.

Water transfers are important in northern New Mexico because they often move water from Indian and Hispanic users to urban and industrial users, but they are only one part of the complex politics of water in the state. All water management initiatives, from permit applications to adjudications, are more suspect than they are in other states because they threaten the complex and fragile web of communal uses or threaten to deprive Indians or Hispanics of control over their future. This case illustrates the subtle, intangible values at stake when water is reallocated in a culturally diverse setting to which the parties bring long histories of past injustices and the difficulties of designing responsive institutions.

Irrigation was first practiced in the upper Rio Grande Valley in A.D. 1000 by the Anasazi, and later by the Pueblo Indians, thought to be their descendants. Spanish settlements on the northern frontier of their empire soon followed, and these also used simple forms of irrigated agriculture that continued when New Mexico passed to the United States in 1848. Each cultural group has modified the natural setting and brought a different perspective to bear on water use. Today, three sometimes intensely conflicting cultural concepts about water coexist in northern New Mexico. To many Pueblo Indians, water is a spiritual force to be respected. The Hispanics have a long tradition of water as a community resource to be shared equitably by all users. Anglo settlement brought the new idea that resources are commodities to be bought and sold.

THE SETTING

The Rio Grande can be divided into three segments: the upper Rio Grande, the Rio Chama, and the middle Rio Grande. Water transfer issues in northern New Mexico focus on the upper Rio Grande Valley (Figure 7.1). A detailed description of the rich physical, cultural, institutional, and economic interactions in the valley is found in *The Upper Rio Grande: A Guide to Decision Making* (Shupe and Folk-Williams, 1988).

Physical Setting

UPPER RIO GRANDE

The Rio Grande River enters New Mexico from Colorado through a steep canyon that makes irrigation canals and other diversions impractical. The river then flows undiminished for about the first 70 mi (112 km) in northern New Mexico, gaining from tributaries and ground

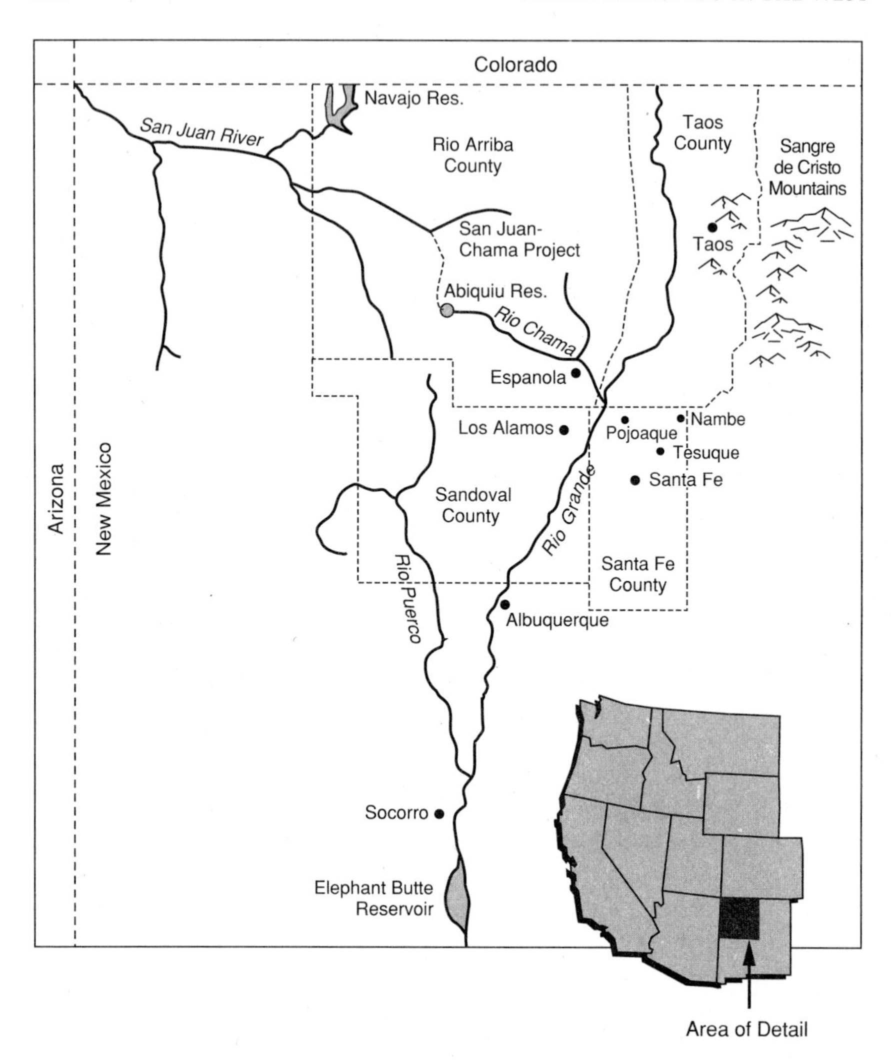

FIGURE 7.1 Main waterways and related features, northern New Mexico.

water. The Rio Grande is used intensively in this reach by rafters, sightseers, and others who find free-flowing water of significant value. A 48-mi (77-km) segment of this portion was designated a Wild and Scenic River in 1968. Below the canyon, north of Espanola, diversion dams and irrigation begin influencing the mainstream of the river. The tributaries of the Rio Grande in northern New Mexico also are used for irrigation. Some of the small farms irrigated by Pueblo Indians and Hispanic acequia associations have histories stretching back several centuries. More recent developments, such as municipalities, ranches, and mines, also have tapped the tributaries of the Rio Grande. Water quality problems and concerns over limited supplies are a result of these increased uses.

RIO CHAMA

The Rio Chama joins the Rio Grande near Espanola, bringing water from the natural runoff of the Rio Chama watershed and water imported from the San Juan River, which is part of the Colorado River basin. This transbasin transfer passes through a series of diversions and tunnels built by the Bureau of Reclamation as part of the San Juan-Chama Project, transporting water through the continental divide to Heron Reservoir. Water from Heron Reservoir enters the Rio Chama, which flows into El Vado Reservoir, built by the Middle Rio Grande Conservancy District to store spring runoff and supplement the supply of district irrigators. Releases from El Vado enter a reach of the Rio Chama designated by the New Mexico Legislature as a Scenic and Pastoral River. This empties into Abiquiu Reservoir, in the area immortalized in artist Georgia O'Keeffe's paintings, and then flows 35 mi (56 km) to join the Rio Grande at Espanola.

MIDDLE RIO GRANDE

Below Espanola, the Rio Grande flows more than 200 mi (320 km) before reaching Elephant Butte Reservoir. The river is highly controlled in this middle reach. The river is first gaged at Otowi Bridge near Espanola, after which it enters the Cochiti Reservoir. The Jemez Reservoir below Cochiti Dam is used solely for sediment control and recreation. Intensive irrigation occurs downstream of Cochiti Dam, the primary users being irrigators in the Middle Rio Grande Conservancy District. Water in this region is also allotted to several Pueblo Indian tribes, as well as to municipal and instream flow uses.

The Rio Grande in the middle reach is so modified by drainage projects, channelization, levees, and other flood control measures that the face and the course of the river have been altered dramatically. Despite these changes, riparian habitat remains along extensive portions of the middle Rio Grande, including one of the world's largest cottonwood groves. The Bosque del Apache and other wildlife refuges provide critical habitat for birds and wildlife.

Institutional and Legal Setting

COLONIAL SPANISH LAW

Hispanic communal irrigation practices originated from Spanish colonization practices and laws. Disputes continue about the exact nature of this Spanish heritage because New Mexico towns claim that Pueblo rights are superior to subsequent Anglo appropriations. But these arguments aside, the basic thrust of Spanish water policy, as set forth in the late eighteenth century Plan de Pitic, was that available water supplies were to be shared among colonists and Indians. In practice, Indian entitlements were limited, but they have always been a legitimate part of New Mexico's water system (Meyer, 1984).

Under Spanish law the right to use water for domestic uses followed from the grant of crown lands. Irrigation was a different matter: the crown reserved the right to grant or withhold irrigation rights with individual land grants because large-scale consumptive water use was a matter of community interest. In New Mexico, water rights could be obtained through an original land grant, through a subsequent administrative grant, or by a later judicial determination. This process, known as composition, could "cleanse, authenticate, and even alter original grants" and could resolve disputes through the application of equitable principles (Meyer, 1984). Voluntary use agreements also were recognized.

Spanish jurisprudence recognized a number of relevant principles that can be found to varying degrees in modern water law. Such factors include title based on a land grant, prior use, need, the protection of third parties, the relationship between the proposed use and the community interest, and legal right and whether the use was by a public community or private individuals. Prior use was important but not controlling. Third parties, which often meant Indians, were protected under Spanish jurisprudence, and third party interests included the crown or state interest (Meyer, 1984).

THE RIO GRANDE COMPACTS OF 1938

The Rio Grande is divided by the Rio Grande Compact of 1938 into two areas below and above Elephant Butte. Transfers can take place relatively freely within each area, but transfers between the two areas are difficult because they would upset the compact balance. The compact allocates the river among three states and between Mexico and the United States. Colorado and New Mexico have rights reflecting their historic uses above Elephant Butte Reservoir under specific streamflow conditions. Water in excess of these uses must flow to Texas and Mexico. If any state wants to increase its diversions beyond the compact level, it must augment the supply in some way.

New Mexico's compact obligations are measured at the Otowi Bridge gaging station, a few miles south of Espanola. When those flows are low, New Mexico must allow 57 percent of the flow at the Otowi gage to pass to Elephant Butte Reservoir. The percentage of Otowi flows that must reach Elephant Butte rises gradually, until they equal 1,400,000 acre-feet (1,726,900 megaliters (ML)). At this point, New Mexico must deliver 71 percent, and the amount of water that may be consumed between Otowi and Elephant Butte reaches its maximum, 405,000 acre-feet (499,570 ML). Above this level the whole increase in Otowi flows must be delivered to Elephant Butte. Thus there is a considerable incentive for New Mexico to consume as much water as possible above Otowi.

THE MIDDLE RIO GRANDE CONSERVANCY DISTRICT

In the central area of the basin is one of New Mexico's oldest conservancy districts, the Middle Rio Grande Conservancy District. Created in 1925, the Middle Rio Grande Conservancy District's distribution and drainage system consists of 202 mi (323 km) of canals (6 mi (10 km) of which are concrete lined), 579 mi (926 km) of laterals (4 mi (6 km) of which are concrete lined), and 399 mi (638 km) of open concrete and pipe drains (Folk-Williams and Hilgendorf, 1982). The district includes more than 123,000 irrigable acres, of which more than 87,000 acres were being irrigated in 1980 (Clark, 1987; MacDonnell, 1990; Shupe and Folk-Williams, 1988). The Middle Rio Grande Conservancy District has zealously guarded water rights appurtenant to land within its service area and has prevented the transfer of irrigation rights initiated since the district's formation. The inability to transfer post-1925 rights out of this district to other uses creates greater pressures to transfer irrigation rights out of other districts.

Economic Setting

AGRICULTURE AND EMPLOYMENT

Northern New Mexico is an economically poor but culturally rich area. People are attracted to the area from all over the country to enjoy the beautiful environment and the living presence of ancient Indian and Hispanic cultures. Agriculture—measured strictly in dollars—is marginal. In Taos and Rio Arriba counties, the growing season is short because of the high altitudes; yields and stocking rates are correspondingly low. In 1986, per capita incomes in this area were low—$7,827 in Rio Arriba and $8,337 in Taos. Unemployment rates were high in 1986—20 and 27 percent, respectively—and in 1979 some 28 percent of the people in the two counties were living below the poverty level (Bureau of Business and Economic Research, 1989). The prime value of agriculture in this region is cultural. It is the basis of a traditional way of life for ethnic communities that are proud of their histories and have a high level of interest in maintaining their historic way of life despite pressures for change.

Northern farms characteristically have low cash sales; annual marketings were below $2,500 on 50 percent of all farms in Rio Arriba and Taos counties in 1987 and below $10,000 on 83 percent. Of the farms in the two counties, 19 percent are less than 10 acres; 67 percent are less than 180 acres. The number of farms grew between 1982 and 1987. These figures indicate a stable, if marginal, agricultural sector that seems to be experiencing moderate improvement in financial viability. Of the farm operators in the region, 62 percent report that their principal occupation is other than farming, but only 34 percent worked more than 200 days off-farm during 1987. These figures indicate a high level of rural underemployment (Nunn, 1989).

Below the Otowi Bridge compact accounting point lie Santa Fe and Sandoval counties. Much of the irrigated area of these counties is north or west of the Middle Rio Grande Conservancy District boundaries. Crop yields are better than in the north because growing seasons are slightly longer and the terrain is more level, but the dollar value of agriculture is low here, too. The upper valley is less rural than the counties to the north. Santa Fe is the state's capital and a world-renowned tourist center and haven for urban refugees, where out-of-state dollars contribute to a relative prosperity. Per capita income in 1986 was $14,047 in Santa Fe and $11,082 in Sandoval; 1986 unemployment rates were 6.3 percent and 8.7 percent, respectively (Bureau of Business and Economic Research, 1989). Farms in these counties have even lower marketings than their northern neighbors: some 56 percent of the farms in the area had sales below $2,500 in 1987, and 81 percent had sales below $10,000. Farms in the area also tend to be smaller: in 1987, 27 percent were less than 9 acres and 68 percent less than 180 acres (Nunn, 1989).

Because the Rio Grande is a continuous hydrologic unit, settlement generally hugs the river. A water market has developed in response to urban and recreational demands. Many of New Mexico's growing towns and cities are located north of Elephant Butte Reservoir—the Albuquerque metropolitan area, Santa Fe, Taos, Espanola, and Socorro in particular. In addition to municipal uses, there are growing recreational uses in this region, especially for ski resorts and recreational subdivisions, but most of the water transaction activity in northern New Mexico is below Otowi Bridge. Santa Fe is the northernmost city drawing water from the Rio Grande basin below Otowi. Albuquerque made acquisition of water rights for future growth a priority in 1979, although actual purchases for private water rights did not begin until 1982. Other buyers in the water rights market below Otowi include utility companies, businesses, and developers, who must bring rights with them as a condition of subdivision approval (Nunn, 1989).

Santa Fe's water is supplied by the Sangre de Cristo Water Company, which has a mixture of surface rights in the Santa Fe River, ground water rights from wells within the city, and contracts for San Juan-Chama Project water. These latter supplies are obtained through wells in the Buckman well fields, which draw from the aquifers associated with the Rio Grande, so that this pumping can be offset by San Juan-Chama Project rights. However, the pumping has local impacts on flows in the Rio Nambe-Pojoaque and Rio Tesuque, and these cannot be offset by San Juan-Chama Project rights. With its current rights on these rivers acquired to offset pumping impacts, Sangre de Cristo is limited to average withdrawals from the Buckman fields of 2,800 acre-feet (3,450 per ML) per year (Nunn, 1989).

The firm yield of water rights owned by Sangre de Cristo will not support demand at levels projected for the year 2000. To increase the water supply from the Buckman fields, Santa Fe will have to expand the field and acquire rights in the Nambe-Pojoaque and Tesuque systems; alternatively, San Juan-Chama Project water could be diverted directly from the Rio Grande, at considerably higher cost. Current estimates are $2.4 million for expansion of the Buckman fields, excluding the cost of water rights acquisition, and $6 million or more for a surface diversion system (Santa Fe Metropolitan Water Board, 1988). Four Indian Pueblos, the Pojoaque Valley Irrigation District, the Tesuque Mutual Domestic Water Consumers Association, 28 community ditch associations, and 2,250 individual water rights holders in the Nambe, Pojoaque, and Tesuque watersheds will have their rights determined in the adjudication. Water transfers, especially transfers that remove water from community ditches, have given rise to local conflicts. Prices of water rights in the area have also been high; there were several sales at around $5,400 per acre-foot ($4,380 per ML) in the early 1980s, when the price in the Albuquerque area was about $1,000 per acre-foot ($810 per ML) (Nunn, 1989).

Although Santa Fe has no immediate plans to purchase additional water rights in the Nambe-Pojoaque and Tesuque systems, it seems likely that it will again be in the market for water rights in this area sometime during the next 40 years. Only rights from this local watershed will release Santa Fe's constraints on pumping from the Buckman well field, and prices for these rights are expected to be high (Nunn, 1989).

Albuquerque relies exclusively on ground water from the Albuquerque aquifer, which is fed by the river. Thus for Albuquerque to expand its pumping, it must consider the effects that expansion would have on the river, because surface water and ground water are managed conjunctively in New Mexico. When the Rio Grande Underground Water Basin was declared in 1956, Albuquerque's vested rights were evaluated at 18,672 acre-feet (15,140 ML) of consumptive use in the Rio Grande. When city pumping affects the riverflow in excess of this amount, it must be offset by the purchase and retirement of surface rights (*City of Albuquerque* v. *Reynolds*, 1963). For many years the city's effect on the river was actually to increase flow levels, because of the time lag between ground water pumping and the release into the river of treated municipal effluent. Not until 1976 did Albuquerque's pumping begin to diminish flows in the Rio Grande. The effects of Albuquerque's pumping should soon require the city to compensate for the additional rights affected. Like other cities, Albuquerque is looking to transmountain diversions to support continued growth. The city will use some of the 48,200 acre-feet (59,460 ML) of San Juan-Chama Project water it acquired in 1963 toward its debt to the Rio Grande. In the period from 2025 to 2060 (based on the present rate of population growth and per capita water use), the city will need at least 45,000 acre-feet (55,510 ML) in addition to its vested rights and San Juan-Chama Project water to offset the effects of its current and future pumping.

Albuquerque has been acquiring rights gradually, based on a standing offer of $1,000 for an acre-foot ($810 per ML) of perpetual right, initiated in 1982. The city has purchased about 1,700 acre-feet (2,100 ML) of water rights in more than 25 separate transactions (Nunn, 1989).

The state engineer estimates that non-Indian irrigators between Cochiti Dam and Elephant Butte have consumptive use rights for about 128,000 acre-feet (160,000 ML). Of these, approximately 65,000 acre-feet (80,180 ML) were rights developed after 1907 within the Middle Rio Grande Conservancy District. (Rights developed before the district was created in 1907 do not require district approval for a transfer.) It is not clear whether the Middle Rio Grande Conservancy District itself holds the property interest in rights developed after 1907 or whether it can lease these rights, some of which are held under state law and some under federal law. The district's current policy is to discourage transfer of these water rights outside its boundaries (Nunn, 1989).

Nevertheless, 63,000 acre-feet (77,710 ML) of the non-Indian agricultural water rights are outside the Middle Rio Grande Conservancy District below the Otowi Bridge Compact accounting point and can be transferred to municipalities under current rules and policies. About 7,000 acre-feet (8,640 ML) has already been purchased, leaving 56,000 acre-feet (69,080 ML) of potentially marketable rights. Albuquerque will need at least 45,000 acre-feet (55,510 ML) to make the offsets required against its well fields in the period 2030 to 2060. In addition, there will likely be an increase in water deliveries in excess of current rights in small municipalities in the middle and southern regions of the Rio Grande of about 14,000 acre-feet (17,270 ML) per year by 2030 (Nunn, 1989). By these figures, it is estimated that demand for 61,500 acre-feet (75,860 ML) or more in water rights will be competing for 56,000 acre-feet (69,080 ML) of pre-1907 irrigation rights and whatever additional contracts for the 8,000 acre-feet (9,870 ML) or so of unallocated San Juan-Chama Project water can be obtained for this area (Nunn, 1989).

THIRD PARTY IMPACTS

Types of Water Transfers

Water transfers involving changes in the place and purpose of use of water rights held under state law have occurred regularly and with relative ease in New Mexico for decades. The state has long had a strong, highly professional state engineer whose goal was to define water rights to encourage the maximum consumptive use of the state's supplies before they flowed to other states and to Mexico. On the average, more than 100 change-of-water applications were filed per year in New Mexico from 1975 to 1987 (MacDonnell, 1990). Fifty-one percent of transfer applications involved less than 10 acre-feet (12 ML), though the average size of a transfer was 91 acre-feet (112 ML) of water. Only 4.5 percent of the applications were protested, and 93 percent of those filed were approved, with 78 percent being approved within 6 months of the date the application was filed (MacDonnell, 1990). New Mexico transfer applications are processed more quickly by the state engineer, involve fewer protests, and involve significantly lower costs in terms of attorneys' fees and consultants' expenses paid by transfer applicants than applications in neighboring southwestern states (Colby, 1990a; MacDonnell, 1990).

In the upper Rio Grande basin, several types of transfers are particularly important because they raise new types of public interest concerns. The first type consists of transfers from traditional small-scale acequia agriculture to recreational uses, such as snowmaking and resort development. The nationally noted *Sleeper* case (*In re Application of Sleeper,* 1985) is a paradigm of a culturally controversial transfer that would be routine in almost any other state. In this widely discussed court decision, strongly reminiscent of the John Nichols novel (and subsequent movie) about water and power in the Taos valley, *The Milagro Beanfield War,* a community's local ditch association protested a voluntary transfer of 75 acre-feet (92 ML) of water in Rio Arriba County to support a ski resort. The transfer was approved by the state engineer after proof that no vested water rights would be impaired, but a district judge voided the approval because the transfer was contrary to the public interest, which was defined as preservation of a local subsistence economy. A Rio Arriba native and doctor of anthropology testified that the development would provide only menial jobs for the local inhabitants and would erode the community's traditional subsistence agriculture (Parden, 1989). There is a tendency to romanticize the virtues of rural poverty, but northern New Mexico has persevered, to the extent that it has, in the celebration of these virtues, and the trial judge held that the public interest required that the preservation of cultural heritage be superior to economic development. However, the decision was reversed on appeal because the court decided that the law in effect at the time did not allow the consideration of public interest in transfers.

New Mexico has since amended its transfer statute to allow public interest review of transfers, in an effort to ward off efforts by El Paso, Texas, to appropriate New Mexico ground water for use in Texas, so the case may be a forerunner of expanded public interest arguments. The focus on the efforts to block El Paso has led to a definition of public interest in terms of the long-term needs of the state for the more traditional uses of water, but there is considerable interest within the state in expanding the definition to include the preservation of ethnic identity. If there is any case that supports including such values as part of public interest criteria, New Mexico presents a truly compelling instance. Some advocates of this view draw an analogy between historic preservation zoning and local water use: they believe that local communities should be allowed to control water use by defining the public interest as the desires of the local community. This view has been endorsed by respected experts in the state (DuMars and Minnis, 1989).

A second type of transfer involves sales of water rights by farmers to urban water providers. Agriculture-to-urban transfers have occurred routinely in the Albuquerque area, as noted earlier, for more than a decade with little controversy. Most of these water rights are being leased back to the irrigators until the water is actually needed by the city. Other cities' water utilities and private developers have also purchased local senior agricultural rights (Colby, 1990b). These transfers have not been perceived as threatening the local agricultural economy and rural communities, probably because the farms are already located in an urbanizing area and farm families in this area rely extensively on off-farm jobs to supplement their farm income. In contrast to the transfers in the Albuquerque area, agriculture-to-urban transfers that have occurred near Santa Fe and Espanola in northern New Mexico have generated local opposition to the removal of water from acequia communities.

A third type of transfer involves the San Juan-Chama Project. The city of Taos has a contract for project water but cannot conveniently take delivery of that water because it is located upstream of where the Rio Chama joins the Rio Grande. Instead Taos takes a quantity of water equivalent to its contract entitlements out of the Rio Grande near Taos. This, and similar exchanges, have prompted concerns about reduced flows and deterioration of riparian plants and wildlife and about reduced water available to acequias diverting from the stretch of the river between Taos and the confluence with the Rio Chama. In addition to having concerns about these exchanges, some northern New Mexico acequia representatives feel that the San Juan-Chama Project made northern New Mexico acequias much more vulnerable to future transfers and prompted the ongoing adjudications that are viewed by some as hostile to acequia interests.

A fourth series of transfer-like events involves the adjudications being conducted throughout the upper Rio Grande basin. Whereas adjudications such as the Santa Fe case discussed above generally are not viewed as water transfers but simply as the confirmation of preexisting rights, in this case they have been characterized as water transfers (or even water theft) by some affected parties and have been perceived as generating third party impacts similar to those experienced when water rights are transferred to new locations or uses.

A fifth type of transfer with potential third party impacts involves change of water rights ownership within an acequia. In many acequias, new residents (often Anglos) have bought land and thus acquired water rights. Although the water often is still used on the same land and for crops similar to those grown before the change in ownership, the character of the acequia may change substantially.

The new Anglo water rights owners often become acequia spokespersons and take over leadership roles. This trend further erodes the unique culture and community cohesiveness fostered by the traditional water use system.

Impacts on Community and Environmental Values

Increasing demands for water that has been used by cultures and communities living in New Mexico for centuries have led to decisions to transfer water and water rights, in some cases between hydrological basins. These transfers have had and continue to have impacts on third parties in the state. These third parties are primarily Indians, Hispanics, and, more recently, environmental uses such as recreation, wildlife, and instream values.

THE ACEQUIA COMMUNITIES

Spanish water use developed around the construction and maintenance of acequias (irrigation ditches). Communities, many of which continue today, developed around the acequias, and irrigators administered the allocation of water in times of shortage through ad hoc negotiations that shared the water on the basis of preexisting shares and needs (Crawford, 1987). The acequias—as both the ditches and the community institutions that managed them came to be known—controlled water allocation until the reclamation era. But they "could not . . . adapt to the large-scale investments necessary for extensive irrigation works on which further growth of the territory must depend" (Clark, 1987). Today, about 700 acequias lie in the northern part of the Rio Grande basin, watering 160,000 acres on 12,000 farms, 70 percent of which are 20 acres or less (Lovato, 1974). Acequias are now both traditional water institutions and contemporary political subdivisions of the state. They have the power to condemn land and to tax members for service and improvements. Under state law, the acequia is controlled by an elected commission and a mayordomo who, like the voters that elect them, must have shares in the ditch. Historically, the shares have been transferable outside the acequia. However, many believe that adjudication changes the acequia's water from a resource held in common to one that is owned privately by individuals.

By most accounts, the water transfers and changes in patterns of water use in the upper Rio Grande basin would be characterized as economically efficient. They reallocate water from low-value crop production or meadow irrigation to more valuable second-home developments, snowmaking, new suburbs, and other uses for which individuals are willing to pay far more for the water than its value for crop production. However, if the less tangible values associated with traditional acequia communities are considered—such as historical significance, cultural preservation, social stability, and contributions to regional uniqueness—there is reason to encourage special considerations when water transfers involving acequias are evaluated by the state engineer's office.

If one wanted to make a case for protecting communities as entities, northern New Mexico would be the example to use. Those who wish to preserve the integrity of acequia communities have argued that water rights allocation should be done at the community rather than the state level. Community advocates often draw an analogy between historic preservation zoning and acequia preservation, noting that zoning could be used in a similar manner to protect cultural areas. Historic zoning preserves the unique character of an area by prohibiting changes in the exterior of all buildings within a district without community approval. Similarly, community control could be used to review water transfers and thus discourage changes that are inconsistent with a community's vision of itself. Water allocation traditionally has been a state, not a local function, so such an approach would require a major revision of the state's water allocation laws.

INDIAN COMMUNITIES

There are 22 Indian reservations within the basin boundaries of the Rio Grande in New Mexico, most in northern New Mexico. These include the Jicarilla Apache Reservation, 3 Navajo communities, and 18 Indian Pueblos. It is mainly the water rights of the Pueblos that are at issue in the Rio Grande. The Pueblo Indians of northern New Mexico are heirs to an indigenous North American irrigation tradition that spread from central Mexico, Meso-America, to the Hohokam of central Arizona, to the Anasazi. "[T]he farmers of the prehistoric Southwest adopted the plant varieties that would survive and produce abundantly in the hot, dry environment and while they stored their crops for time of need, they also enhanced the success of their agricultural activities by using irrigation" (Hunt, 1987). By the time of the Spanish discovery and conquest of Mexico and the modern Southwest, the New Mexico Pueblos had developed a settled irrigation-based existence. Sophisticated land tenure arrangements existed, but they differed from the postfeudal concepts of private property brought to North America by the English. Land was not an alienable commodity but a community or family, often matriarchical, resource. In Anglo-American terms, clans and individuals acquired the right to use specific plots rather than ownership of land.

Indian water rights are equally complicated, although for different reasons. Indigenous practices are the basis for Pueblo water rights, but the form is the product of Anglo jurisprudence. This is a source of frustration to some tribal leaders. In addition, the Pueblos see water both as a commodity to be developed and traded and as a spiritual resource. Water rights of several Pueblos are being adjudicated. They clearly have rights, but federal court decisions have determined that the *Winters* doctrine (described in Chapter 3) is not the only basis for Pueblo water rights because of a history of Spanish and Mexican recognition of the Pueblo Indians' settled way of life in permanent villages. Title to Pueblo land has always been legally different from that in other Indian communities. These prior titles were recognized by the United States when it took over control of Mexican territory under the Treaty of Guadalupe Hidalgo in 1848.

Pueblo Indian water rights are complicated because of this history. In an ongoing adjudication north of Santa Fe (*New Mexico* v. *Aamodt*, 1976), the court developed a two-tiered system of Pueblo water rights. Pueblo lands held before U.S. sovereignty hold aboriginal water rights. Lands acquired afterward carry with them *Winters* rights. To complicate matters, the court held that the Pueblo Lands Act of 1924 also "fixed the measure of pueblo water rights to acreage irrigated as of that date." If the concept of this case is upheld on appeal, the Pueblo Lands Act must be interpreted to determine the upper limit placed on Pueblo water rights by Congress. The adjudication has established that Pueblo water rights with aboriginal priorities do exist, although the doctrinal basis differs from general federal reserved Indian rights. Pueblo leaders have expressed some frustration with the adjudication process. They do not object to the substance of the rights declared but to the fact that the process is non-Indian.

INSTREAM FLOW

This case study focused almost exclusively on the consumptive use of water by Hispanic and Indian users. There is an environmental dimension to these claims because they are generally directed at preservation of the status quo. Northern New Mexico does have traditional environmental problems. The area has high-quality instream flows as well as stresses on rivers used for recreation. Endangered species protection has not played as large a role in northern

As the Rio Grande River enters New Mexico from Colorado, it flows undiminished for about 70 miles (112 km). This reach is used by rafters, sightseers, and others. Farther downstream, north of Espanola, diversion dams and irrigation begin influencing the mainstem of the river. CREDIT: Todd Sargent, University of Arizona.

New Mexico as in the Truckee-Carson or Yakima basins, but water management will have to accommodate this use in the future.

New Mexico, almost alone among the western states, has not joined the movement to adopt new laws to protect instream flows. Legislation to protect streamflows was introduced in the New Mexico legislature several times in the 1980s, but no laws were passed. However, there have been nonlegislative case-by-case efforts to maintain instream flows in important reaches of the Rio Grande basin of New Mexico. White water rafting groups have negotiated with Albuquerque and the Middle Rio Grande Conservancy District to enhance recreational opportunities on the Rio Chama. The city and district have agreed, on a year-by-year basis, to schedule their summer releases from El Vado Reservoir to correspond with weekend rafting needs in the stretch above Abiquiu Reservoir.

The Bureau of Reclamation has also cooperated over the past several years by extending the San Juan-Chama annual water delivery deadline from December 31 to April in order to allow for more consistent flow to maintain winter fisheries. The bureau has also recognized Indian religion and culture by agreeing to release a minimum natural flow from Nambe Falls Dam to ensure that 0.5 cubic foot per second is maintained at Nambe Falls. Federal agencies are also working to protect or enhance instream flows by claiming rights in court based on federal law. Other recent studies document the continued importance of reliable streamflows to the New Mexico economy, which relies heavily on outdoor recreation and tourism (Ward, 1987). Endangered species protection for fish in the San Juan and Delores river basins may require instream flow designations by New Mexico. The matter is being pursued as a part of broader efforts to make the Animas-La Plata Project acceptable from an endangered species perspective and to implement the southern Colorado Ute water settlement passed by Congress.

CONCLUSIONS

Northern New Mexico teaches two important, related lessons for future water management. First, the promotion of efficiency—as it has been defined by federal and state water policies—can have detrimental impacts on low-income, culturally different users. These users are at risk when water administrators seek to maximize the economic value of water by project development or water marketing. Second, it is not easy to incorporate heightened sensitivity to social and economic inequities into the structure of existing water allocation institutions. Study of water transfers in northern New Mexico makes clear the need to broaden the definition of public interest to acknowledge the value of water to community cohesiveness, along with the more typical factors considered. The Hispanic community of northern New Mexico, for example, considers itself a third party affected by almost every water decision—determinations of water rights, transfer of a parcel of land from one owner to another, and changes in the place or use of a water right. State water law considers the public interest in new appropriations and transfers, but fails to include in such considerations the traditional community values tied to water.

An expanded definition of public interest is crucial because Hispanics have scant resources to protect their interests in water transfers affecting them. They also must confront daunting challenges in trying to remedy past shortcomings in adjudications that often overlook their rights. Small water transfers in New Mexico, in these circumstances, face acute and complex challenges that would be unlikely to arise in most other states.

New Mexico has now enacted legislation prohibiting the transfer of a water right from one use or place to another where the effect would be "detrimental to the public welfare or contrary to the con-

One Perspective on Adjudications: "Dancing for Water"

"[E]veryone has been startled at least once by the miraculous imagery of the saying that water can be made to flow uphill toward money. It is perhaps part of the genius of the image that for the moment at least it seems to settle the arguments that bring it into play. Everything has its price. We live in a world of buying and selling. We all know that. What else can we say?

"Yet we also live in a world of other kinds of value, and sooner or later we will need to ask whether it is good for anything but money that water can be made to flow uphill, whether it is bad for streams, rivers, traditions, communities, and even individuals.

"Several weeks ago I sat in on a water rights adjudication hearing in Santa Fe—my first, so I was able to pretend to view it through somewhat innocent eyes. There was the small, cramped courtroom in the federal courthouse, probably not unlike courtrooms and hearing rooms where these proceedings have droned on in New Mexico for hours and days and months and even years and decades. There were the twenty-five or thirty silent, patient defendants, Questa landowners, working men, probably miners or former miners, Hispanic all, dressed in plaid western shirts and cowboy boots. There were the attorneys and legal assistants for the Office of the State Engineer, four in all, Anglos all, with their table covered with huge aerial photographs, yellow legal pads, clipboards of lists, notebooks. There was the plaintiffs' comparatively empty table, presided over by a lone attorney from Northern New Mexico Legal Services and a volunteer assistant from Questa. There was a tripod from which hung detailed maps of the section of Questa being debated, with parcels of land blocked out in green and red and pink. There was the imitation wood panelled box of the witness stand where sat, during my time there, a succession of Mr. Raels and Mr. Valdezes. There was the judge, a kindly black-suited gentlemen who patiently elicited and presided over a river of minutiae about fences and culverts and walls and houses and ditches and pastures and the placement of this or that boulder, all of which made up the substance of the proceeding.

"What I witnessed for a few hours was the operation of that legal mechanism by which water is prepared for its eventual pumping toward money. It has to be adjudicated, it has to have its claims of ownership documented, it has to have its title quieted, it has to be made merchantable, saleable, which is what enables it to be freed up from land, acequia, community, and tradition. . . .

"Water should go—we all know—to those who tend it, who use it, who love it, who dance for it, and it should flow downhill from stream to river, and river to sea. Yet we have licensed our society to scheme for it to flow toward those with money and power, and we have turned over our public servants to them, we have in effect given them the keys to the treasury so that no expense will be spared in drawing lines, making maps, conducting research, making surveys, and filling vaults and basements with mountains of legal testimony.

"What is oddest of all—I thought as I regained the open sky of a fretful March afternoon—is that all this legalistic hairsplitting over water rights needs the long-term ratifications of the weather in order to work at all. And the weather, of course, is what no one can ever bring to court. Being, I suppose you would say, above the law. Capable, in short, of rendering the vast social labor of adjudication quite irrelevant—in these times of rapid climatic change all over the globe. A kind of higher adjudication of the environment, you might say, that could be trying to tell us that fooling around with the elements is something we should think and talk long and carefully about before anything else, and in ways that make us all more neighborly, not less so."

SOURCE: Crawford (1990).

servation of water" (New Mexico Statutes Annotated § 72-12-7 (replaced 1975)). The statute was enacted in an effort to ward off attempts by El Paso, Texas, to appropriate New Mexico ground water for use in Texas, but it could be the basis for expanded public interest review of intrastate as well as interstate transfers. The focus on the efforts to block El Paso has led to a definition of public interest in terms of the long-term needs of the state for the more traditional uses of water, but there is considerable interest within the state in expanding the definition to include the preservation of ethnic identity. Questions remain, however, about whether a fair determination of the public welfare can be made in the administrative and judicial arena (DuMars and Minnis, 1989) and whether this decision should be made at the state or local level.

If there is a compelling case illustrating an instance in which community identity can be seen as a public interest value, it is New Mexico. This case shows how a combination of economic and noneconomic values supports the preservation of the traditional water allocation system. It adds weight to the argument that public welfare interests can best be determined at the local level by a regional water planning process (DuMars and Minnis, 1989). Such an approach could, in the end, expedite transfers by building public confidence in the fairness of evaluation procedures and by reducing uncertainty about what standards apply.

REFERENCES

Bureau of Business and Economic Research. 1989.

City of Albuquerque v. Reynolds, 71 N.M. 428, 379 P.2d 73 (1963).

Clark, I. 1987. P. 100 in Water in New Mexico: A History of Its Management and Use. Albuquerque: University of New Mexico Press.

Colby, B. G. 1990a. Enhancing instream flow values in an era of water marketing. Water Resources Research 26(6):1113-1120.

Colby, B. G. 1990b. Transactions costs and efficiency in western water. American Journal of Agricultural Economics 72(5):1184-1192.

Crawford, S. 1987. Mayordomo. New York: Doubleday.

Crawford, S. 1990. Dancing for water. Journal of the West 32:265-266.

DuMars, C. T., and M. Minnis. 1989. New Mexico water law: Determining public welfare values in water rights allocation. Arizona Law Review 31:817-839.

Folk-Williams, J. A., and L. Hilgendorf. 1982. What Indian Water Means to the West. Santa Fe, N. Mex.: Western Network. 153 pp.

Forrest, S. 1989. P. 3 in The Preservation of the Village: New Mexico's Hispanics and the New Deal. Albuquerque: University of New Mexico Press.

Hunt, R. D. 1987. Indian Agriculture in America: Prehistory to the Present. Lawrence: University of Kansas Press.

In re Application of Sleeper, No. RA 84-53 N.M. Dist. Court for Rio Arriba County (1985).

Lovato, P. 1974. Las Acequias del Norte. New Mexico State Planning Commission, Taos, N. Mex.

MacDonnell, L. 1990. Shifting the uses of water in the West: An overview. In L. MacDonnell, ed., Moving the West's Water to New Uses: Winners and Losers. Boulder: University of Colorado, Natural Resources Law Center.

Meyer, M. 1984. P. 134 in Water in the Hispanic Southwest. Tucson: University of Arizona Press.

New Mexico v. Aamodt, 537 F.2d 1102, 10th Cir. (1976).

Nunn, S. C. 1989. Trading Conserved Water: Market Incentives for Agricultural Conservation. Albuquerque: University of New Mexico, Department of Economics.

Parden, S. 1989. The Milagro Beanfield War revisited in Escenada Land and Water Association v. Sleeper: Public welfare defies transfer of water rights. Natural Resources Journal 29:861.

Santa Fe Metropolitan Water Board. 1988. Long Range Water Planning Study for the Santa Fe Area, Phase I Report. Santa Fe, N. Mex.

Shupe, S. J., and J. Folk-Williams. 1988. The Upper Rio Grande: A Guide to Decision Making. Sante Fe, N. Mex.: Western Network.

Ward, F. A. 1987. Economics of water allocation to instream uses in a fully appropriated river basin: Evidence from a New Mexico wild river. Water Resources Research 23:381-392.

8

The Yakima Basin in Washington: Will Transfers Occur Without Judicial or Legislative Pressures?

The Yakima River basin is a rich agricultural area located in central Washington State on the east slope of the Cascade Mountain range (Figure 8.1). The Yakima River is a major tributary of the Columbia River, annually producing an average of about 3.4 million acre-feet (4.2 million megaliters (ML)) of streamflow. The Yakima and the Columbia once supported one of the world's great salmon runs, which was shared among the region's Indian, sport, and commercial fishermen. Today, salmon throughout the Northwest and in the Yakima basin are under great stress, and both Indian tribes and environmentalists are pressing for substantial restoration of historic runs.

The basin, with its familiar pattern in which irrigated agriculture is asked to use water more efficiently so new demands can be accommodated, is one that would seem amenable to water transfers. But for a number of reasons, pressures for reallocation have not been great enough to induce significant use of these measures. Thus one lesson that can be drawn from the Yakima case is that markets will not reallocate water from agricultural to other uses, such as tribal or environmental, without a substantial judicial or legislative change in the allocation rules, backed by an infusion of public or private resources.

Land ownership in the Yakima basin is approximately 40 percent federal, 30 percent Indian, and 30 percent private. Within 3 years of the Reclamation Act of 1902, the federal government began construction of Tieton Reservoir and purchased the private-investor-owned Sunnyside Valley Irrigation Project. In 1906, Congress authorized the Yakima Project. Over the following 20-year period, six storage reservoirs (with a capacity of 1,000,000 acre-feet, or 1.23 million ML) and 2,000 mi (3,200 km) of conveyance canals were built. Many uses developed for the waters of the basin, but the principal diversionary use is the irrigation of about 500,000 acres of land. Crops grown on these lands in 1986 had a value of $383 million. The principal agricultural products were 135,000 acres of forage crops, 46,000 acres of grains, 95,000 acres of fruits, 27,000 acres of vegetables, and 38,000 acres of field crops.

Historically, Yakima River basin streams sustained six large anadromous fish runs. Each year, as many as 500,000 adult sockeye, Chinook (spring, summer, and fall runs), and coho salmon and steelhead trout returned to the river system after their migration to the Pacific

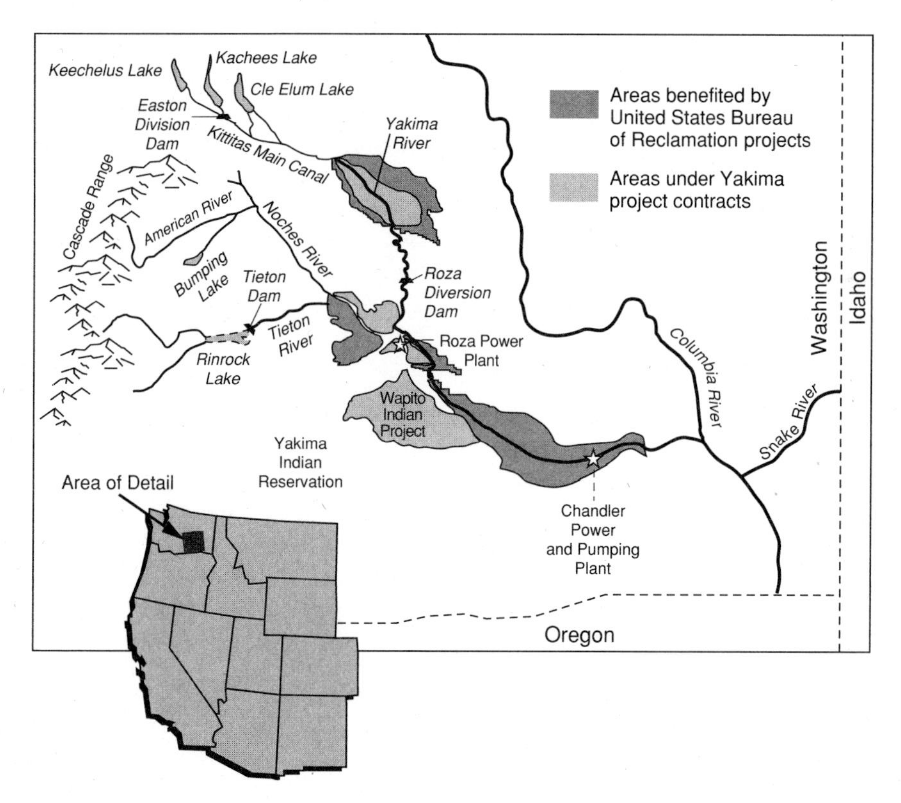

FIGURE 8.1 Main waterways and related features, Yakima valley, Washington State.

Ocean. In the 1970s a record low was hit, however, when fewer than 1,000 fish returned. The decrease has been caused by the cumulative impacts of human activities, including commercial and recreational fish harvest, construction and operation of Columbia River main stem dams, construction and operation of Yakima basin water storage reservoirs and irrigation water delivery systems, diversion of water from the Yakima basin streams, bypasses for hydroelectric power production, and general habitat degradation.

Restoration of the Yakima basin will not be easy. The water of the Yakima River and its tributaries has been fully appropriated since the beginning of the twentieth century; present rights to water actually exceed the supply during most years. The water is used productively. As a result, there are severe conflicts between those who divert it for uses such as irrigation and those who seek to restore the anadromous fish runs and preserve instream flows during low water years. The Yakima basin is a basin with intense competing demands, a shortage of water to satisfy those demands, and little institutional pressure for reallocation. There are conflicting water needs for irrigation (both Indian and non-Indian), hydroelectric power, instream fisheries (both resident and anadromous), recreation, and municipal and industrial purposes. In addition, a senior treaty-based Indian claim for water that would allow expansion of present uses and restoration of fishery flows raises the issue of involuntary reallocation.

How to provide adequate flows and enhancement facilities to protect and restore anadromous fish is a central issue in the Yakima basin. Unlike the other sites highlighted in this report, there is little municipal pressure to bid water away from agricultural use. Environmental needs alone are not enough to create pressure for transfers because instream uses are not recognized in the state's water rights system, although the state can set minimum flows.

THE SETTING

The Pacific Northwest is looked upon by many as an area of water abundance. With the Columbia River discharging approximately 180 million acre-feet (220 million ML) to the ocean each year, some would question whether there could ever be a use for "all that water." But as is the case throughout the West, there is more demand for water in the Yakima River basin than the resource is able to sustain over time. Conflict among uses is particularly critical during below-normal water years.

Within the basin is a potentially volatile mix: a rich and productive agricultural area and an Indian tribe—the Yakima Indian Nation—with latent water rights. The large Yakima Indian Reservation (some 1.3 million acres) is located in the southern part of the basin and is occupied by a confederation of tribes who originally inhabited about 12 million acres. The enrolled membership of the tribe is 8,940, of whom approximately 60 percent live on the reservation. The Treaty of 1855 ceded title of about 90 percent of their land to the United States and created the reservation so that the tribe could become, inter alia, an agricultural people. The treaty reserved many rights within the ceded area, including rights to hunt, fish, and gather berries, roots, and plants at "all usual accustomed places." Historically, members harvested salmon from the Yakima River by traditional scaffolds constructed along the stream bank.

Irrigation in the basin commenced in the latter part of the nineteenth century, after the treaty with the Yakima Nation. In 1905 the Bureau of Reclamation began construction of water storage at Bumping Lake, Keechelus, Kachees, Cle Elum, and Tieton reservoirs. There are also three hydropower projects—Chandler, Roza, and Wapatox—that divert streamflow year-round. While these facilities contribute to the uses of the water supply, they also add to the conflicts.

In 1977, the driest year experienced in the basin in modern memory, conflicts between the Yakima Nation and irrigators led to an effort to reopen a water rights decree (*Kittitas Reclamation District et al.* v. *Sunnyside Valley Irrigation District et al.*, 1977) that had governed the use of water for agricultural purposes since 1945. This decree provided that the parties were to have equal rights with respect to priority in the delivery of water from the river system except for certain uses. These equal rights were declared to be proratable and were found to be subject to pro rata diminution during periods when the total water supply available was less than required for their satisfaction. The excepted uses were the early irrigation canals that existed in the basin prior to the start of the construction of the irrigation project by the Bureau of Reclamation. The rights for these canals were found to be nonproratable and superior to the proratable rights. In times of extreme water shortage, parties with nonproratable rights sometimes receive part of their water from storage in federal reservoirs even though some of these beneficiaries did not participate in repayment of the costs of construction of the reservoirs. This situation is a source of some contention within the basin.

In the 1977 challenge the court ruled that the decree was being interpreted correctly by the Bureau of Reclamation, which had been given responsibility to administer its terms. This ruling led the Yakima Nation to file an action in federal district court (*Confederated Tribes and Bands of the Yakima Nation* v. *United States et al.*, 1977) to determine the priority and quantity of its water rights. The state of Washington also filed an action that same year in state court (*State of Washington* v. *Acquavella et al.*, 1977) for a general adjudication of water rights in the basin. The Yakima Nation's federal court action was deferred to the jurisdiction of the state court under provisions of the McCarran Amendment (43U.S.C.666).

Part of the Yakima Nation reservation (about 130,000 acres) is irrigated by diversion of approximately 650,000 acre-feet (800,000 ML) of water each year from the river system. Supply of about half (305,000-acre-feet, or 376,000 ML) of this water is nonproratable, whereas the remainder is subject to proration. The United States, for the Yakima Nation, has claimed a total of 1,030,505 acre-feet (1,271,000 ML) of water from the river for irrigation and some 1,250,000 acre-feet (1,542,000 ML) for instream flows as part of the *Acquavella* general adjudication. These claims were made under the *Winters* reserved water rights doctrine (*United States* v. *Winters*, 1908) and as "time immemorial" rights, but the court has defined the tribe's rights very narrowly.

In a partial summary judgment issued on July 17, 1990, the presiding judge in *Acquavella* ruled that the Yakima Nation's water rights "[h]ave been fulfilled and limited by a combination of federal reserved and Washington state law water rights," and thereby denied the tribe's claims to additional diversionary rights. The court further found with regard to the Yakima Nation's instream flow claims that "[t]he reserved treaty rights shall be no more than those quantities of water necessary for the maintenance of fish and aquatic life in the Yakima River and its tributaries as and to the extent that such fish and aquatic life exist at this time and no more."

A motion for reconsideration filed by the United States was granted, and on November 29, 1990, the court entered an amended partial summary judgment as final judgment. The court found that the Yakima Nation had a water right for fish with a "time immemorial" priority date but that the right had been diminished to "the minimum instream flow necessary to maintain fish life in the river." It was further stated, "In view of ever changing circumstances, it would be inappropriate for the court to set specific, discrete quantifications." An appeal to the Washington State Supreme Court on behalf of the tribe is likely.

The U.S. claims for instream flows have been made in amounts up to 2.1 million acre-feet (2.6 million ML) annually to satisfy anadromous fish needs, or about two-thirds of the average annual water supply. These flows are needed for upstream passage of returning adult fish, spawning and egg incubation, juvenile rearing and outmigration of "smolts" (young salmon returning to the ocean), to stop or reduce sedimentation and siltation problems, and to reduce high water temperatures in the lower river. Until the claims of the Yakima Nation are quantified, it may be difficult for any other entity to assert enough water rights to effect an increase above the present instream flows sufficient to protect and enhance anadromous fish production. Resolution of these claims will likely influence the federal fishery agencies, the U.S. Fish and Wildlife Service, and the National Marine Fisheries Service, since the United States was a party to the 1945 decree.

There have been a variety of federal and state legislative attempts to resolve the irrigation, tribal, and fisheries conflicts in the basin, but none has yet succeeded. In 1979, however, Congress authorized the Yakima River Basin Water Enhancement Project (P.L. 96-162, 93 Stat. 12411, 1979), and this effort is making some progress. This project has a number of purposes:

- to provide supplemental irrigation water to currently irrigated lands,
- to provide water for new irrigation on the Yakima Indian Reservation,
- to increase instream flows for aquatic life (anadromous fish),
- to develop a comprehensive plan to manage the basin's water resources, and
- to settle the water rights disputes raised in the *Acquavella* case.

Authorization of this project has been followed by a number of studies and efforts by study teams, policy groups, and roundtable groups to negotiate and agree on a plan for action. Some significant steps have been taken, but there has not been full agreement on some measures considered, particularly the need for and expected benefits from agricultural water conservation measures and the construction of new storage reservoirs. Environmental groups have urged that an aggressive conservation program be the first step in the enhancement program before any new storage is built.

With the passage of the Pacific Northwest Electric Power Planning and Conservation Act (P.L. 96-501, 16 U.S.C. 839, 94 Stat. 2697, 1980), funding became available through the Bonneville Power Administration (BPA) for off-site enhancement of anadromous fish habitat and construction of some fish passage facilities at existing diversion dams and fish hatcheries. More than $55 million worth of such facilities have been constructed with BPA funding as part of Phase I of the Yakima River Basin Water Enhancement Project.

In 1989 the state of Washington passed a law (R.C.W. 90.38) specific to the Yakima basin with the expressed purpose "to work with the United States and various water users of the Yakima River Basin in a program designed to satisfy both existing rights, and other presently unmet, as well as future, needs of the basin." This law creates a class of "trust water right," which is defined as "that portion of an existing water right, constituting net water savings, that is no longer required to be diverted for beneficial use due to the installation of a water conservation project that improves an existing system." The Washington Department of Ecology is empowered to enter into contracts to provide funding for water conservation projects and to, thereby, acquire a percentage of the water conserved to be used for "instream flows and/or irrigation use."

No projects have been built under this new law, but it is hoped by the parties that it will aid in promoting water conservation in the basin and will result in a water supply that the state may reallocate to provide or enhance instream flows. A major limitation to the usefulness of the law is the "no injury" provision, which states that, "[n]o exercise of a trust water right may be authorized unless the department first determines that no existing water rights, junior or senior in priority, will be impaired as to their exercise or injured in any manner whatever by such authorization" (State of Washington Law, 1989).

Officials of the Bureau of Reclamation and fishery interests generally believe that improved management of the existing water supply can increase fishery production within the Yakima system. This would be accomplished by controlling flows through certain critical reaches during the spawning, rearing, and migration periods of the year. An interesting example of the management potential using the flexibility currently available in the system is the "flip-flop" operation of storage and release through the Tieton to the Naches and upper Yakima spawning reaches. Because the fish spawning runs are earlier in the season in the Naches River than in the Yakima, the storage and release patterns from the two arms of the basin are managed to encourage the salmon to spawn low in the channel and to allow for higher flows to be delivered to protect the eggs from being dewatered before the young fry hatch and emerge from the gravel. Consequently, early in the season, water is delivered to downstream irrigators from the upper Yakima, and flows are held back in the Naches. Later in the season, this pattern is reversed, with delivery coming from the Naches and flows from the upper Yakima held back.

In another attempt to enhance the fishery through water management, the diversion of water to and the operation of federal facilities at the Roza and Chandler power plants were voluntarily subordinated to instream flow requirements by the Bureau of Reclamation starting in 1990 to ease the stress on anadromous fish during low-flow periods. When the flow in the river is less than the amount desired for instream uses, no water will be diverted strictly for power generation purposes. The bureau is investigating other management options as well.

Interests in the basin are moving ahead with Phase II of the Yakima River Basin Water Enhancement Project by installing fish screens at the intakes of a number of smaller irrigation canals. Legislation is being drafted to implement a water conservation project. Efforts are being made to reach consensus on the construction of additional basin water storage, including modification of the gates at Cle Elum Dam; a pipeline to convey runoff for storage in Kachees Lake, which would otherwise be unregulated; and other system improvements. Proponents have not given up on the idea of enlarging the storage capacity of Bumping Lake. This idea is being strongly opposed by environmentalists, at least until all avenues to improve present irrigation use efficiency have been explored.

The Northwest Power Planning Council approved a project that will provide a complex of hatcheries and "acclimation ponds" that are expected to boost salmon and steelhead runs by 85,000 fish. Fish will be reared in ponds containing water from specific natural streams and then will be released into those streams. It is hoped that the surviving fish will return to the streams to spawn naturally.

THIRD PARTY IMPACTS

Large numbers of formal transfers of water and water rights are not currently being proposed in the Yakima River basin. Rather, considerable effort is being made to increase water use efficiency, particularly for irrigation, as a means of freeing up water to reestablish instream flows. But the retention by the state of the no injury rule in the passage of its new law concerning water efficiency and trust water will be an impediment to transfers that might be proposed in the future.

The primary third party impacts illustrated in this case study affect the environment (anadromous fish and wetlands) and indigenous peoples (the Yakima Indian Nation). If the recent partial summary judgment decision in *Acquavella* is not reversed on appeal, the holders of the existing water rights will continue to control the basin, and there is little likelihood that they will be sympathetic to efforts to reallocate supplies to Indians or to support fisheries. The decision provides certainty and stability to the agricultural economy of the basin, but it does little to enhance the restoration of fishery resources.

Some would argue that the anadromous fish are doomed anyway because of actions outside the Yakima basin, such as downstream dams on the Columbia River and fish harvest policies. Why invest in in-basin facilities such as fish screens and hatcheries, they argue, if the smolts cannot pass through the slack water pools of four large federal dams on the lower Columbia River after they negotiate all the barriers on the Yakima? Others have suggested that the only recourse is to list the salmon and steelhead as threatened or endangered species under the Endangered Species Act of 1973. Such a listing would require development of a recovery plan that could bring substantial changes in the operation of federal dams and reservoirs in the basin. It could also affect downstream facilities and operations and fish harvest policies. If there is no clear, objective, and reasonable expectation to restore the fish runs to their former size and importance, it is questionable whether they should be protected at some minimum level of survival.

There is also circumstantial evidence that enhancements to the anadromous fish runs can be achieved through actions within the Yakima basin independent of any major improvements in fish passage problems in the main stem of the Columbia River. After record low numbers of returning fish in the 1970s, increases were observed up to 1986. After that, three drought years with consequently poor flow conditions during the critical out-migration period pushed adult returns down again. However, 1990 was a good flow year (primarily due to subordination of hydropower uses to instream flows), and if the 1992 run of adults is back up near 1986 numbers, it will be evidence that the flow situation in the Yakima basin is one of the major factors limiting fishery production.

The Yakima Nation also wants to improve its irrigation system on the reservation and desires to market unused water off the reservation. Again, the summary judgment in *Acquavella* would limit the tribe to present uses. There would be no opportunity for expansion of irrigation or marketing unless they would be willing to forgo present uses. The state's Saved Water Act (State of Washington Law, 1989), while allowing water transfers from more efficient water use to instream uses, may also cause negative impacts on the environment, particularly native vegetation and wetlands associated with present irrigation systems.

Three potential water transfers could have an impact in the Yakima basin. Two of the potential transfers would constitute a reallocation of the existing basin water supply from traditional uses to instream use. The third would transfer senior rights from annual grain and hay crops to the expanding perennial fruit and nut crops. The potential transfers important to this case are as follows:

1. Agricultural water that might be transferred to the Yakima Nation for anadromous fish protection and enhancement. This would be an involuntary transfer that might be brought about by either state adjudication or judicial findings. The threat of this involuntary transfer is driving efforts to create a cooperative system of management and to improve efficiency, leading to state and potentially federal legislation.

2. Agricultural water that might be transferred to state-held "water rights trust" to enhance anadromous fisheries. This potential transfer would involve voluntary transfers promoted by state legislation to allow "conserved" water from improved irrigation efficiency to be reallocated to other uses. Major concerns are the lack of incentives for landowners to improve their on-farm efficiency because they realize no immediate gain. There are potential losses of wetlands and riparian habitats associated with fixing leaky irrigation conveyance systems. State or federal cost-sharing incentives will likely determine the extent to which these transfers occur.

3. Agricultural water in the superior nonproratable rights that might be transferred to new agricultural use or to firm up the supply to farmers currently holding only proratable water rights. Such transfers are now allowed only during temporary situations. This transfer would be purely voluntary and is largely dependent on the outcome of state adjudication and on incentives to encourage increased efficiency of use by allowing farmers to put the "conserved" water to use on new land or to sell this portion of their right to another user. Legislation would have to clarify these points with respect to waste versus conservation. This transfer would involve less water than the other two transfers.

CONCLUSIONS

The prior appropriation doctrine in use in the West and in the Yakima basin of the state of Washington is a major impediment to open public interest reallocation of water resources because the doctrine creates vested property rights that, pursuant to constitutional guarantees, cannot be taken without just compensation, and creates few incentives for reallocation when there are no large urban areas or national environmental groups to bid for the water. The doctrine does, however, provide the certainty and stability of supply that are important in an irrigated agricultural economy. This feature is particularly valuable in the Yakima basin where the crops grown include major areas of perennial plants such as orchards and vineyards.

Management considerations in the Yakima basin are becoming more broad based. This is caused primarily by the threat of the Yakima Nation's reserved water rights, which would become senior to present rights, as well as the acknowledged water use inefficiencies inherent in the present facilities and water use practices. The system could be made more efficient, and this improvement would free up water that could be held in trust by the state for reallocation to instream uses. However, improvement in water delivery efficiency would occur at considerable cost, and the general public beneficiaries of such improvements (taxpayers, public interest groups, fishermen, and so on) have seldom indicated a willingness to pay. There are, of course, examples of water efficiency improvements that have been financed by municipal interests for the water saved. If currently used water is to be reallocated to other uses in this basin, it appears that the public must find a way to pay for the improvements necessary to make that possible or effect it through the higher courts.

Incentives could be created to encourage improvements in water use efficiency that would then free up large volumes of water for other uses. Such incentives might include

- financial assistance programs,
- changes in law to promote and reward conservation efforts, and
- changes in state water transfer law to make it responsive to market demand.

The Yakima River basin is an area where flow augmentation from outside the basin may be a viable tool to mitigate the effects of present in-basin uses. As indicated, the Columbia River annually discharges about 180 million acre-feet (220 million ML) of water to the Pacific Ocean. There may still be opportunities to divert some of this water into the Yakima.

Laws and regulatory processes also need to recognize and protect instream flows and the environment of an area in the public interest. Officials charged with administration of resources and protection of the public trust must aggressively exercise their authority. It may not be within the purview of state administrative officials to operate outside the limits of statutory law. The public trust doctrine has been adopted as part of the state's water law by courts in California, Idaho, and North Dakota, but what role it will play in reallocation remains uncertain. One method to reallocate water may be for federal and/or state authorities to buy out marginal agricultural lands and retire the associated water rights for dedication to instream uses.

Some of the solutions for the problem of anadromous fish protection and enhancement in the Yakima basin may be gained from the results of petitions seeking endangered species listing for four species of Snake River sockeye and Chinook salmon. Those petitions will result in biological reports on the status of the fish runs and the possible development of recovery plans. They may bring about requirements for modification of dams and reservoirs and their operation to protect the species.

The Bureau of Reclamation can play an important management role within the confines of federal and state law by promoting multiple uses the existing supply of the hydrological system. In addition, the Yakima River Basin Water Enhancement Project can be offered as a viable example of comprehensive river basin planning aimed toward more efficient management of existing water supplies. However, it is not likely to succeed in meeting the multiple use goals within the basin without substantial investments in water conservation and new storage, as well as intensive management on a basinwide scale by the Bureau of Reclamation.

The Yakima basin is unique among the cases studied in this report in that the most likely water transfers are from agriculture to different types of agriculture and from agriculture to instream flows. Furthermore, the motivation behind these potential transfers would be different. Shifts from annual crops (primarily grains) to perennial crops, along with restoration of the anadromous fish runs, necessitate an emphasis on efficient use, reallocation, and management to produce more benefits with existing supplies.

REFERENCES

Confederated Tribes and Bands of the Yakima Nation v. United States et al. United States District Court (1977).

Kittitas Reclamation District et al. v. Sunnyside Valley Irrigation District et al., Civil No. 21, United States District Court (1977).

State of Washington v. Acquavella et al., No. 77-2-01484-5, Superior Court of Yakima County (1977).

State of Washington Law, Chapter 90.38 RCW (1989).

United States v. Winters, 207 U.S. 564 (1908).

9

Central Arizona: The Endless Search for New Supplies to Water the Desert

Central Arizona was settled initially as a mining and agricultural oasis in the desert, but the area is now part of the fastest-growing, most water-short, urbanized state in the West (Figure 9.1). A 1991 survey of growth in Arizona found that "even in the depths of a real estate crisis in parts of the region, jobs and people continued to increase. . . . It has become the new California, a place where palm trees and the desert still beckon dreamers" (Johnson, 1991). Its water allocation laws and policies are premised on the desirability of sustaining the high rate of growth that began after World War II. The Arizona case study illustrates the problems that a water-short state faces in pursing a water policy premised almost exclusively on supply augmentation in the postreclamation era. This chapter looks at the role water transfers are playing in supporting the region's growth. It also examines the problems that have arisen because of the state's approach to water management and looks briefly at new legislation related to ground water recharge and transfers between basins.

Before 1980, Arizona mined its ground water reserves, secure in the belief that the federal government would eventually pipe a share of the Colorado River into the state's interior. In 1980, Arizona implemented a major new water policy, the Ground Water Management Act, designed to achieve four objectives: (1) ground water conservation, (2) the gradual conversion of agricultural areas around Phoenix and Tucson (and their associated water supplies) to urban use, (3) completion of an aqueduct to import Colorado River water,

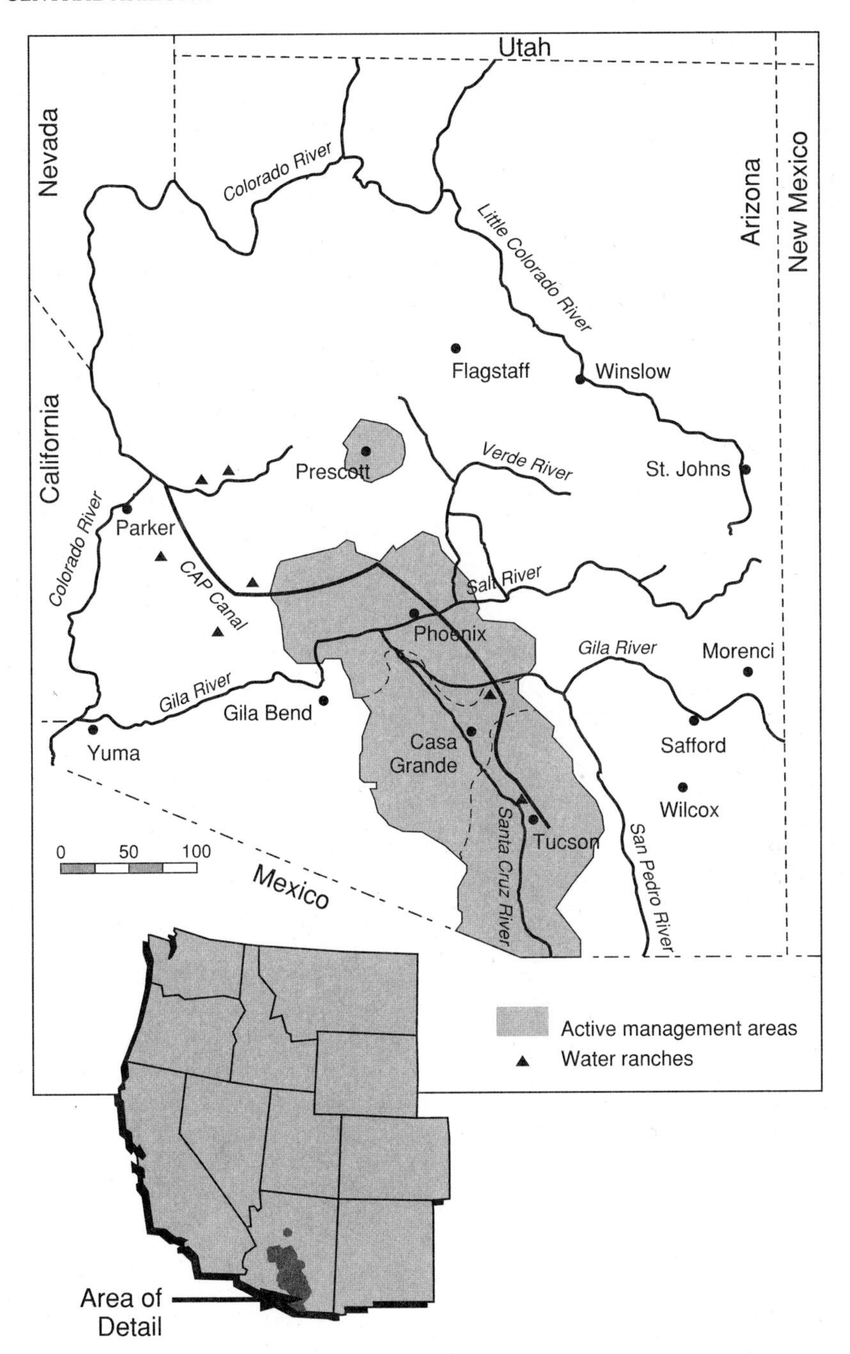

FIGURE 9.1 Main waterways and related features, Arizona.

and (4) removal of state court constraints on transfers of ground water. Arizona law requires no state approval of proposed ground water transfers, although surface water transfers must be reviewed. There are no statutory procedures or substantive protection for area-of-origin interests or other parties who may be affected by a transfer but do not own water rights. Moreover, Arizona law does not recognize any connection between ground water pumping and streamflows, so there is no protection against the adverse impacts of ground water transfers on riparian areas and surface water rights holders.

The Ground Water Management Act's mandate to achieve safe yield in ground water use by the year 2025 and the requirement that new real estate developments demonstrate a 100-year water supply created strong incentives for water transfers. It encouraged cities to reach beyond the boundaries of the urban "active management areas" (AMAs) to acquire and transfer water from rural agricultural areas. In effect, the act encouraged cities to buy farmland solely for its appurtenant water—so-called water farms (Table 9.1). Water farming is extremely controversial. Many people see it as a solution to Arizona's water supply problems because it provides the means to transfer the state's limited water supplies to areas of greatest need. Others, however, are extremely concerned about the adverse effects of water farming, especially on the people and environments of areas of origin (Checchio, 1988).

THE SETTING

Arizona's population is concentrated in the interior desert lowlands, where there is a substantial imbalance between dependable water supplies and the amount of water consumed. The state's growth is a testimonial to the ability of the human species to circumvent the limitations of the natural environment. State water policy is premised on the assumption that supplies can and must be found to meet the demands of continued population growth, especially in the state's two metropolitan centers, Phoenix and Tucson. Because of the high level of urbanization and associated commerce and industry, agriculture's relative contribution to the state's economy has declined sharply. Agriculture now contributes only about 2 percent of the state's employment, while it uses more than 80 percent of the state's water.

Arizona relies on both surface and ground water supplies. Surface flows originating in the highlands of the Mogollon Rim and the White Mountains were used by the Hohokam in prehistoric times along the lower Gila and Salt rivers. Mormon settlers revived the

TABLE 9.1 Arizona Water Farms

	Owner	Location (county)	Acres	Acre-feet per Year (ML per year)[a]	Price ($ millions)	Date Purchased
McMullen Valley	Phoenix	La Paz	14,000	30,000 g.w. (37,000)	$30.5	1986
Avra Valley	Tucson	Pima	22,518	60,000 g.w. (74,000)	$24.73	1971-1986
Pinal County Farms	Mesa	Pinal	11,607	29,918 g.w. (36,900)	$29.14	1985
Planet Ranch	Scottsdale	La Paz	8,400	13,500 s.w. (16,650)	$11.6	1984
Crowder-Weiser Ranch	American Continental	La Paz	7,685	50,000-60,000 g.w. (61,600-74,000)	[b]	1985
Lincoln Ranch	Lincoln Ranch Ltd. Partnership	La Paz	1,040	5,200 g.w. 6,300 s.w. (6,400, 7,770)	$ 5.0	1984
APS	Arizona Public Service Co.	La Paz	12,550	[b]	$ 9.0 (book value)	1980

[a]ML = megaliters; g.w. = ground water; s.w. = surface water.
[b]Unavailable.

SOURCE: Woodard et al. (1988).

The Central Arizona Project has the potential to divert 1.2 million acre-feet (1.5 million ML) of Colorado River water to Phoenix and Tucson. Originally conceived of to bring water for irrigated agriculture, by the end of the 1970s its capability for supplying urban and industrial users was paramount. CREDIT: Todd Sargent, University of Arizona.

irrigation economy in the Salt River valley after the Civil War. The Mormons colonized northern Arizona between 1873 and 1876, settled in the Salt River valley in 1877, and then began to farm along the Gila. The settlements were close to the state's major mining districts, which the Mormons helped provision (Piremen, 1982).

As was the case with the Truckee-Carson Irrigation District in Nevada, the first projects taken on by the new Bureau of Reclamation in the early 1900s in the Salt River Valley involved the rescue of ailing irrigation projects. The bureau's construction of Roosevelt and other dams on the Salt River stabilized the valley's agricultural economy (Fradkin, 1981). In 1917 the bureau transferred ownership of the distribution facilities to the Salt River Project, which delivers water and markets hydroelectric power to the Phoenix area. Other major surface uses are the Welton-Mohawk Irrigation Project lands along the Gila River above Yuma with water diverted from the Colorado, and the San Carlos Irrigation Project along the Gila on the San Carlos Indian Reservation above Phoenix.

Even after the Salt River Project captured almost the entire flow of the Salt drainage, surface supplies were inadequate to support the continued high water consumption required for growth in the Phoenix metropolitan area. Colorado River water was unavailable for financial and legal reasons, except to users along the mainstream, largely near Yuma; only the federal government could have afforded to pay the immense construction and pumping costs of moving Colorado River water to the metropolitan areas. In addition, Arizona shares the lower Colorado River with California, Nevada, and Mexico. Under federal law, Arizona is entitled to an "equitable share" of this interstate stream, but the entitlement was not clearly defined until 1963, when the court limited Arizona to 2.8 million acre-feet (3.5 million megaliters (ML)) per year of the lower basin's allocation of 7.5 million acre-feet (9.3 million ML) per year. Existing diversions from the Colorado in the Yuma area are about 1.1 million acre-feet (1.4 million ML). Arizona's unused Colorado entitlement has always been available to California and, to a lesser extent, to Mexican users. Completion of the Central Arizona Project (CAP), diverting Colorado River water, will give the state the capacity to transport 1.2 million acre-feet (1.5 million ML) of Colorado River water to Phoenix and ultimately to Tucson.

Until the 1980s the CAP was viewed as the complete answer to Arizona's water needs. While waiting for the project's completion, the state made up the difference between its 2.5 million acre-feet (3.1 million ML) of dependable surface supplies and water demand by mining ground water. By the end of the 1970s, the state was running an annual ground water overdraft of between 2.2 and 2.5 million acre-feet (2.7 and 3.1 million ML). The original strategy of CAP proponents was to guarantee the supplies necessary to continue the expansion of irrigated agriculture, but by the end of the 1970s the purpose of CAP had shifted to the bailout of urban and industrial users. Because of the full range of competing demands for CAP water—agriculture, municipal and industrial, and the satisfaction of Indian claims—the project is no longer the complete answer to the state's water needs. The CAP will reduce but not eliminate the need for ground water mining. Thus, in the long term, transfers will still be necessary to support the continued growth of the state's urban areas.

Water transfers have a long but limited history in Arizona. Water transfers occurred as early as 1948, when the city of Prescott bought farmland in the nearby Chino Valley and developed a well field to pump water for domestic use. Controversy arose even then, with local farmers charging that the Prescott pumping exceeded the amount normally needed for agriculture and led to water declines in the basin (Woodard, 1988). Some additional water farming activity took place in the early 1970s, when Tucson began buying and retiring farmland in the Avra Valley, located about 15 mi (24 km) away. More than 21,000 acres was bought for a total purchase price of $22.7 million (at an average price of just over $1,000 per acre) (Woodard 1988).

Although even these early water farm purchases caused concern in their areas of origin and led to numerous lawsuits, they did not create the intense statewide controversy that bloomed when it became clear that the Ground Water Management Act would limit pumping and the CAP would not be a complete substitute source of water to support unlimited growth. These early purchases differed from recent ones in several ways (Woodard 1988):

- the land was relatively nearby;
- the water was in the same hydrologic basin, although in different subbasins;
- the water transfers were limited in scope and driven by a relatively immediate need for water;
- the land was in the same county, so property tax impacts were internalized; and
- the cities incorporated the purchased land into their service area, ensuring an adequate future water supply for local residents.

WATER INSTITUTIONS

Arizona water law is the product of the historic assumption that water should not be a limiting factor in the state's economic development. This assumption is reflected in the appropriation doctrine under which surface rights were recognized, as well as in pre-1980 Arizona ground water law, which did not recognize hydrologic surface-ground water relationships and thus allowed virtually unrestrained ground water pumping. Ground water was not subject to even the mild conservation limitations inherent in the doctrine of prior appropriation (Leshy and Belanger, 1988).

Today, surface water is administered under the appropriation doctrine through the Department of Water Resources (DWR). The DWR also administers the Ground Water Management Act of 1980, the act that ultimately placed the state's immense ground water resource under management in the metropolitan areas of Tucson and Phoenix and in Pinal County.

The Ground Water Management Act drove the state's transfer activities by allowing formerly prohibited transfers with few restrictions. Prior to 1980, Arizona followed the common law of ground water. Transfers are limited under common law doctrines, but in 1976 the Arizona Supreme Court went further (*Farmers Investment Company* v. *Bettwy*, 1976) and held that ground water could not be used off the land from which it was pumped if other users of the same aquifer were injured. The decision "created a storm of protest from the strong Arizona mining lobby" (Kyl, 1982). Industry, working together with cities, succeeded in reversing the decision in the 1980 act. Because the act did not establish any mechanism for evaluating, conditioning, approving, or disapproving such transfers, an unregulated market in rural water as a source of population and economic expansion for urban Arizona was created, and a controversy of historic proportions was generated.

This approach is consistent with Arizona's tradition of water policy decisionmaking. Throughout Arizona's history, the state's water decisions have typically been made outside of formal policymaking institutions and, instead, by direct negotiation among the powerful water-using interests (Gregg, 1990). It is an approach that began early in the state's history, when settlers developed the Salt River valley, even before the 1902 Reclamation Act. This and other early projects set the model for federal reclamation policy: federal money and state control. And, in reality, control was provided by the state's most influential water users.

The Salt River Project (SRP) was created in 1917 to repay the nonreimbursable costs of the project, which had been built by the Bureau of Reclamation, and has become a major supplier of water and energy in the state. Another local institution, the Central Arizona Water Conservation District (CAWCD), was created to handle the contracting and repayment obligations for the CAP. The CAWCD may or may not accrue as much power as has been exercised by the SRP. The Bureau of Reclamation, with its large investment in the state, including the SRP, the Welton-Mohawk Project, the CAP, and the huge Colorado storage project, Lakes Mead and Powell, is a major presence in the state system, but it does not possess the power that one might expect given the scale of its projects. Crucial operating and contracting power was lost to the two local agencies, and the bureau's power to allocate the Colorado—except in times of severe, sustained shortage—is constrained by the law of the river and newly emerging environmental-energy conflicts. Likewise, the Gila River above the confluence with the Salt River near Phoenix is regulated by the San Carlos Irrigation Project, a federal Bureau of Indian Affairs (BIA) project. The BIA is not, however, a major participant in state water politics; off-reservation agricultural water users are major beneficiaries of the project.

When agriculture was the state's most powerful interest, Arizona refused to restrict the use of ground water by owners of overlying lands. Agriculture was strong enough to resist pro rata pumping cutbacks suggested during the negotiations that led to the Ground Water Management Act. Collectively, irrigators have powerful property interests in federally developed surface water to complement the state-created grandfathered ground water rights recognized by the state legislature in 1980. Municipalities, as agents for the state's growing urban population, are now the dominant force in Arizona water politics, but they have carried forward the style of brokered deals to support the dominant use of water.

The concentration of the state's population in two metropolitan areas confers unusual power on those who supply water to these areas. Water service in Phoenix is bifurcated between the Salt River Project, which has access to both surface and ground water, and non-SRP areas. The city of Phoenix holds substantial surface water rights, but most non-SRP metropolitan areas are dependent on ground water rights and CAP water. Concentrating economic and political power in Phoenix has not led to centralized, comprehensive management. Instead, it led to an unstable combination of ad hoc joint actions and frequent competition among independent municipalities, especially such major suburbs as Scottsdale, Tempe, and Mesa.

Because water providers have a responsibility to provide long-term assured water supplies, they tend to perceive the highest value of water to lie in its role as a critical component in economic growth. This view distresses other water interests concerned with water as a community value in areas of origin or as an ecological, recreational, and scenic amenity. Supply augmentation has been the gospel of the state, and transfers are simply a modern means of augmenting supply since the federal government will no longer build big water supply projects. Historically, the entire political apparatus has tended to support supply augmentation as the basic water policy even when doing so may not have been in the rational self-interest of some participants. The sustained and disciplined support for the CAP for more than 20 years, even as its purpose shifted from supplemental irrigation to municipal and industrial supply, is in the tradition of unified support for supply development in the Salt River Valley that has dominated water policymaking there in the last century.

Circumstances forced the major players to supplement this policy with conservation in 1980. Arizona was forced to conserve its ground water when Secretary of the Interior Cecil Andrus, on behalf of the Carter administration, refused to support funding for completion of the CAP until the state took steps to reduce its ground water pumping. Governor Bruce Babbitt used the funding denial to force the three major water user interests—urban, agricultural, and mining—to agree to the most stringent ground water conservation regime in the nation, the Ground Water Management Act of 1980.

Basically, mining and urban interests struck a deal with agricultural interests about the rate and price of the transition from agricultural to urban use. In brief, agriculture was forced to yield its long-held claim to ownership of all underlying ground water and to accept a stringent short- and long-term conservation regime. Water was shifted from agriculture to urban development, but agriculture was able to capture much of the value of ground water by authority to sell its rights for transfer to urban areas. The 1980 management plans for the Phoenix and Tucson AMAs included support for intensive watershed management (brush removal) and cloud seeding as augmentation measures, despite hostility from mountain communities, outdoor recreationists, and summer-home owners, and uncertain scientific and legal research. In short, the state still relies on distributive politics, which depends on consensus among the large user interests about how the resource will be used to the exclusion of other, more diffuse or less politically powerful interests (Martin et al., 1988). State political leaders have accepted the need for the CAP as an article of bedrock faith in face of the decline of its original beneficiary, agriculture, and serious criticisms of the efficiency of the project (Martin et al., 1988).

Radical as the Ground Water Management Act was, it did not break the general pattern of policymaking by brokered agreements among traditional elites. The act was brokered by a small group of representatives of agriculture, mining, and urban users; the cities' spokespersons were not representative of a diverse range of environmental, social, and economic concerns, but of supply augmentation to ensure unconstrained growth. Farmers were protected by clearing title to water and authorizing transfers; rural community interests were not broadly represented. The resulting agreement was presented to the state legislature on a "take it or leave it" basis, and the legislature took it. The relative power among the traditional interests has shifted some over time, but the number of players at the table has not greatly expanded.

THIRD PARTY IMPACTS

There are five separate major transfer regimes in Arizona: (1) Colorado River water, (2) non-Colorado River surface water, (3) ground water, (4) sewage effluent, and (5) Indian reserved water rights. Transfers in these first three areas, as driven by the Ground Water Management Act, are discussed first, followed by discussion of impacts in the fourth and fifth regimes.

Surface and Ground Water Transfers and Their Social and Environmental Effects

The Ground Water Management Act of 1980 reflects a series of conscious water allocation choices to a much greater degree than is the case with most state water allocation legislation. The act mandates the goal of elimination of the ground water overdraft by the year 2025. Overdraft will be reduced by a series of 5- and 10-year plans promulgated by the state's Department of Water Resources that apply to the state's most populous areas and to the center of its remaining irrigated agriculture. The act divides the state into three areas: (1) Phoenix, Prescott, Tucson, and Pinal County are designated active management areas, and the first three of these are subject to the no overdraft goal; (2) nonagricultural uses are only about 2 percent of Pinal County's consumption, so the goal in this cotton- and citrus-growing area is to preserve the agricultural base as long as possible; and (3) three other agricultural areas are frozen in size by being designated as irrigation nonexpansion areas (INAs). Outside of AMAs and INAs, there is little management or regulation of ground water development. Ground water use is controlled by the pre-1980 common law reasonable use doctrine, but general stream adjudication on the Gila River—which was begun to quantify Indian reserved water rights—may become a vehicle to subject non-AMA ground water to the appropriation doctrine and to integrate ground and surface rights (Leshy and Belanger, 1988). For example, in 1988 the presiding judge issued an order defining appropriable water as including subsurface flows that support surface ones.

Colorado River water is available to Arizona as a result of congressional legislation and a U.S. Supreme Court decision (*Arizona* v. *California*, 1963) that confirm the state's equitable share of this interstate resource. The linchpin of Arizona's ground water use policy has been reliance on the CAP, a tax-subsidized vehicle for pumping water 2,000 ft (667 m) uphill from the Colorado to central Arizona. Outside of the Colorado and Salt rivers, surface use and thus surface transfers make up a minor portion of the state's total water consumption. Arizona is entitled to 2.8 million acre-feet (3.5 million ML) of Colorado River water, of which 1.2 million acre-feet (1.5 million ML) are to be delivered by the CAP. In 1968, however, Arizona had to subordinate the CAP to the claims of California and the upper basin to get the federal government to pump the water over the mountains to Phoenix and Tucson. Paradoxically, the availability of CAP has spurred interest in ground water transfers because it provides the physical means to convey the water.

Rural communities consider themselves especially at risk from the lack of review of transfers from nonregulated areas to AMAs because the 1980 law encourages transfers. When water farming first began in 1948, the water farms were adjacent to the place of intended use and the original understanding was that the cities would confine themselves to adjacent agricultural areas. Hence, no effort was made to address the consequences in remote rural areas. Water farm purchases in the late 1980s triggered intense controversy because of the large scale of the land purchases and the potential economic, social, and environmental costs. Water farming in western Arizona is possible because the CAP canal is a means of transporting water to Phoenix during dry years if space is available. Use of the canal for ground water transport has not been completely resolved.

In the three urban AMAs the basic purpose of the act is to manage the conversion of agricultural land to urban uses through land transfers. The act places the major burden on agriculture to conserve water and creates economic incentives for the conversion from agricultural to urban use. Transfers are permitted under restricted conditions. The underlying assumption is that there would be a close correlation between the source of the water and the urban use. Existing wells may continue to pump if they qualify for one of four vested rights. The right must be a grandfathered irrigation right, a service area right, a state withdrawal permit, or a storage and recovery permit. In brief, agricultural land that was retired from production in anticipation of urban use can obtain an "irrigation grandfathered right." The right is based on historic pumping patterns, but historic water duties will shrink over time as progressively more stringent conservation measures kick in.

To use the water for urban purposes, grandfathered rights must be converted to Type I rights. To qualify for a Type I right, land must be permanently withdrawn from farming. The maximum amount of ground water that can be pumped is 3 acre-feet (3.7 ML) per acre per year. Type I rights must be withdrawn from the land originally associated with the right, but they can be transferred. Type I rights may be transferred within AMA subbasins, but rights acquired after 1986 cannot be transferred out of an AMA. Type I rights can be transferred from outside an AMA to land within an AMA in the same basin or subbasin even if other pumpers suffer damage. Inter-subbasin transfers require the payment of damages. Intra-AMA transfers were made more complicated in an effort to confine water providers to their service areas. Rights may be acquired and transferred between subbasins subject to the limitation that transportation in excess of 3 acre-feet (3.7 ML) per year requires the payment of damages.

The major incentive in the Ground Water Management Act to transfers is the requirement that no new land development may occur unless the developer can guarantee adequate supply to meet anticipated needs for at least 100 years. To calculate this supply, only the amount of water that could be pumped when the AMA reaches safe yield can be counted as a guaranteed supply. Although CAP contracts may be added to this base, CAP water counts only until the year 2001. As a result, the Ground Water Management Act created strong incentives for cities to purchase water farms and to transfer water from non-AMAs to AMAs because it linked new development to imported water.

The Ground Water Management Act removes many barriers to transfers and exchanges, but it leaves many third parties vulnerable to the costs of transfers, especially environmental and area-of-origin values. Transfers of both surface and ground water, which are generally hydrologically related, can have serious environmental consequences. Transfers and exchanges can stress or eliminate riparian ecosystems, which exist in limited numbers in the state and many times contain rare and threatened species.

For example, some cities, such as Prescott, cannot receive CAP water because they are too far from the CAP aqueduct. However, they are seeking to exchange their CAP rights for other surface rights held by other interests. Such exchanges may affect the minimum flows of streams. The riparian system may adjust to the reduction in minimum flows by migrating toward the stream channel with recruitment of young riparian trees, while the width of the riparian band is reduced through loss of trees on the edges of the floodplain. If these adjustments reach an equilibrium, the availability of riparian habitat will not drastically change. But adjustments do not take into account periodic spring floods, which may destroy the inwardly migrating riparian community. The ultimate consequence may be a narrowing or perhaps loss of the riparian band, with little or no recruitment to compensate for these losses. The reduction or loss of the riparian band reduces wildlife habitat, threatens stream bank stability, and degrades downstream water quality.

Transfers can be designed to offset this loss if they provide instream flow protection. Arizona is one of the few states that allow private instream appropriations. In 1983 the state approved an instream flow appropriation by an environmental organization and has since approved three more, but the application process appears to be on hold, with more than 70 applications pending as of early 1991. Instream flow appropriations are, at best, a minor means of protecting riverine ecosystems because these rights are always junior to all existing consumptive rights. However, given the scarcity of surface flows, instream rights may also be acquired by the purchase of existing senior consumptive rights. The Public Trust Doctrine, although not yet in use in Arizona, may be considered in the future to protect riverine ecosystems on public lands.

The land surface also can be damaged by transfers. When water is transferred away, wetlands in the area of origin may be drained and farmland retired. When farmland is retired, as in the Avra Valley west of Tucson, the vegetative cover changes. Fallow fields attract weeds such as Russian thistle (tumbleweed) that then blow onto neighboring lands. Dust and erosion also can increase unless a proper substitute vegetative cover is established.

Water farms have been identified as possible causes of economic and social disruption. There is a large gap between the law and the expectations of both urban and rural residents. A survey of community leaders in rural areas found most of them believing that waters originating in their watershed belong to their area, although the bedrock principle of prior appropriation is that the right to use water is not tied to the watershed of origin. Further, they believe that communities will suffer substantial losses if water is removed and that these losses cannot be compensated by fiscal transfers (Oggins and Ingram, 1990).

One of the most controversial examples of environmental damage by water farming is the city of Scottsdale's purchase of the Planet Ranch in La Paz County. The ranch pumps water from a large shallow aquifer supplied by the Bill Williams River and raises thousands of acres of crops. Scottsdale has had plans to increase the water reserves originally held by the Planet Ranch by acquiring an additional 75,000 acre-feet (92,510 ML) of excess flood flows from Alamo

Lake upstream, which currently recharge the aquifer. To avoid abandoning its water rights during the interim period while it is arranging transport of the water by the CAP canal or another conduit, Scottsdale must continue to use the water. Irrigated acreage has expanded in order to perfect more water rights. As pumping has increased, the riparian forests adjacent to the ranch have died; forests downstream in a U.S. Fish and Wildlife Service refuge are stressed except where beavers have created ponds. The manner by which the water is transported to Scottsdale is critical. If the water is piped out of the Bill Williams River, streamflows will continue to decrease. However, if the city uses the Bill Williams River to transport water back to the Colorado River and then into the CAP aqueduct, streamflows in the Bill Williams River and associated riparian values could be preserved.

In 1988 and 1989 the state tried to resolve some of the rural-versus-urban conflicts by designating some rural ground water basins as urban reserves (or "sacrifice areas," as seen by rural residents) and withdrawing others by designating them as environmentally sensitive areas. This effort failed in the legislature. It is clear that a balance between equity and efficiency has yet to be struck in Arizona.

These water transfer issues remain a part of the legislative agenda in Arizona. Area-of-origin communities have strong reasons for attempting to revisit water transfer policy. They would like compensation for tax losses and economic losses, and a share of the profits made in the sale of water to higher-valued urban uses. The most urgent desire, however, is for assurance that sufficient water will be left to sustain reasonable levels of economic and population growth in rural areas (Gregg, 1990). Rural people are questioning why the AMAs, in order to reach "safe yield" use rates of their own ground water aquifers, are permitted to exhaust aquifers elsewhere in the state.

Although these equity arguments seem strong, there is no real incentive for urban interests to yield the advantages they have from the present law. In 1989 the Arizona legislature killed legislation that would have prevented or limited transfers from most rural areas but would have allowed them from nine specifically designated basins, six of which were in La Paz County. In 1990 the legislature ignored area-of-origin concerns again and proposed a "ground water replenishment district" for the Phoenix AMA.

Finally, in 1991 the legislature joined supply augmentation and area-of-origin protection. A regional management authority was created for the Phoenix area, which can levy taxes to replenish ground water. The authority must use the taxes to replenish CAP water before it searches for additional supplies. To protect areas of origin, the legislation imposes a ground water transportation fee schedule ranging from $3 to $30 per acre-foot and permits "voluntary" payments to the area of origin in lieu of property taxes. The state still has not confronted the issue of whether safe yield can be achieved under the current policies or whether more stringent measures such as land retirement will be needed (Glennon, 1991).

Effects of Transfers of Sewage Effluent

The Arizona Supreme Court has made effluent a major source of transferable water. In 1989 the court held that treated effluent is a separate source of water from surface and ground water and that it may be sold by municipalities, over the objections of downstream water rights holders, subject only to the beneficial use requirement. Thus, although cities have no duty to continue to discharge effluent, this court decision may preclude the use of a stream for carriage to contract purchasers because the court suggested that existing appropriators may use the water among themselves (*Arizona Public Service Co.* v. *Long*, 1989). Still, effluent will become the major source of golf course irrigation and Indian water rights, especially as the U.S. Environmental Protection Agency and the Arizona Department of Environmental Quality set higher water quality standards for effluents released into surface streams subject to the Clean Water Act. This may cause municipalities to seek other outlets for effluent use. Unfortunately, if effluent has been the sole perennial source of instream flow, as in the Salt River below the Phoenix wastewater treatment plants, riparian values are likely to suffer if the effluent is put to other uses. The loss of this water source may be extremely detrimental to certain riverine systems. Studies are needed to evaluate this on a case-by-case basis.

Effects of Transfers of Indian Water Rights

Arizona has many Indian reservations, both on the Colorado River and in central Arizona, close to Phoenix and Tucson. Collectively these Indian tribes assert the largest block of *Winters* claims. Arizona, along with other western states, shares a tendency to envision apocalyptic consequences of the fulfillment of Indian claims. In practice, Indians must either seek federal or state assistance to turn paper water claims into wet water or must sell or lease the water to non-Indian users. Thus tribes have some incentive to accept far less than their maximum entitlement in return for capital to initiate development projects. Indian water rights are in various stages of quantification. Many remain unadjudicated, but some tribes have obtained congressionally negotiated settlements. As more water is dedicated to the reservations, economic incentives are created for the Indians to trade this water for cash. Whether or not Indian water rights can be transferred off the reservation is unclear. It appears that such transfers will require congressional consent. The Tohono O'odham and Ak-Chin settlements have sanctioned the transfer of Indian CAP water to municipalities, and these will undoubtedly influence future legislation.

CONCLUSIONS

Arizona illustrates an aspect of the water transfer issue that is not present in many of the other studies. In the Truckee-Carson basins, the Imperial Valley, the Central Valley, the Yakima basin, and Colorado cases, increased transfers would be an efficient response to the water demands of urban areas or environmental values. Further, in some of these areas, existing laws and allocation institutions may unduly impede transfers. The situation in Arizona is the opposite. It is not clear that large-scale rural-to-urban transfers represent the most efficient and fair allocation option open to the state. Existing laws and institutions unduly favor this option to the exclusion of other means of meeting the state's water demands and at the expense of unrepresented third party interests.

The state's long history of preferring supply augmentation at any cost to any form of conservation has shaped the use of water transfers. Instead of a break with past allocation practices in the name of fairness and efficiency, water transfers represent the continuation of past allocation practices. Arguments on both sides of the water transfer debate in Arizona can be compelling. Indeed, the current free market approach does promote the most economically efficient use of water resources by allowing water to move into urban areas willing to pay high prices for it. But this leaves the state's rural populations and other less powerful, diffuse values at risk and without a strong voice in the decisionmaking process. And the question remains: Is the market approach the best way to allocate the state's scarce natural resources, particularly one as critical as water? Clearly, there is a public interest or community value in water that must be protected. Arizona lacks the policies and mechanisms needed to facilitate a full discussion of the merits and problems of water transfers among all affected parties. The state's historical style of water resource decisionmaking—which relies on private negotiations among major interests—acts as a disadvantage in addressing this issue equitably and with long-term vision.

As this report was being drafted, changes occurred in Arizona's approach to water management. Having recognized some of the problems arising from the Ground Water Management Act of 1980, the Arizona legislature in 1991 passed two pieces of legislation that attempt to improve the state's water transfer process. The Ground Water Transportation Act (S.B. 1055) represents a compromise between rural and urban interests. For rural areas the measure protects water supplies for economic growth, whereas for urban areas it protects the ability to obtain supplemental water to assist in demonstrating an assured water supply. In general, the bill restricts transfers of ground water from rural ground water basins to AMAs, with some exceptions. The bill determines the quantities of ground water available to AMAs that can be used to demonstrate an assured water supply and sets fees or voluntary in lieu payments to be made by AMAs for water transfer.

The Ground Water Replenishment Act (H.B. 2499) was passed in conjunction with the Ground Water Transportation Act. It authorizes the establishment of a ground water replenishment district for the Phoenix AMA. The ground water replenishment district may lease wells from irrigation districts for transfer; however, limits are set on the depth of ground water withdrawal and the rate of decline. It is anticipated that surface water within the AMA will be used for ground water replenishment when ground water transfers take place.

REFERENCES

Arizona v. California, 373 U.S. 546 (1963).

Arizona Public Service Co. v. Long, 160 Ariz. 429, 773 P.2d 988 (1989).

Checchio, E. 1988. Water Farming: The Promise and Problems of Water Transfers in Arizona. Issue Paper No. 4. Tucson: Arizona Resources Research Center.

Farmers Investment Company v. Bettwy, 113 Ariz. 520, 558 P.2d 14 (1976).

Fradkin, P. L. 1981. A River No More: The Colorado River and the West. New York: Alfred A. Knopf.

Glennon, R. J. 1991. Because that's where the water is: Retiring current water uses to achieve safe-yield objectives of the Arizona Groundwater Management Act. Arizona Law Review 33:89-114.

Gregg, F. 1990. The widening circle: The Groundwater Management Act in the context of Arizona water policy evolution. In Taking the Arizona Groundwater Act into the Nineties, the Proceeding of a Conference. Tucson: University of Arizona, Udall Center for Studies in Public Policy and Water Resources Research Center.

Johnson, D. 1991. Arid economy, desert states thrive. New York Times (May 13):A1, C3.

Kyl, J. L. 1982. The 1980 Arizona Ground Water Management Act: From inception to current constitutional challenge. University of Colorado Law Review 53(3):471-476.

Leshy, L., and J. Belanger. 1988. Arizona law where ground and surface water meet. Arizona State Law Journal 20(3):657-748.

Martin, W. E., H. M. Ingram, D. C. Cory, and M. G. Wallace. 1988. Toward sustaining a desert metropolis. In Water and Arid Lands of the Western United States. New York: Cambridge University Press.

Oggins, C., and H. Ingram. 1990. Does Anybody Win? The Community Consequences of Rural-to-Urban Water Transfer: An Arizona Perspective. Issue Paper No. 2. Tucson: University of Arizona, Udall Center for Studies in Pubic Policy.

Piremen, B. 1982. Pp. 179-180 in Arizona Historic Land. New York: Alfred A. Knopf.

Woodard, G. C. 1988. The Water Transfer Process in Arizona: Analysis of Impacts and Legislative Options. Tucson: University of Arizona, College of Business and Public Administration.

10

California's Central Valley: Fear and Loathing in Potential Water Markets

California faces ever-increasing prospects of water scarcity and thus needs to find new ways to meet a wide range of demands. Both strong public preference for the environmental amenities associated with free-flowing water and the escalating financial costs of developing new water supplies have made it increasingly difficult for the state to pursue its past policy of building large dams and canals to meet new needs. At the same time, population growth in California is approximately 750,000 per year, and projections indicate that this growth will continue beyond the turn of the century (California Department of Water Resources, 1987a).

Since it appears unlikely that the state will embark on major new water development schemes, some reallocation of existing supplies is inevitable. One major reallocation is already under way as the Metropolitan Water District salvages approximately 100,000 acre-feet (123,350 megaliters (ML)) annually from the Imperial Irrigation District (see Chapter 11). Other reallocation schemes include court-ordered reductions in diversions by the city of Los Angeles from the Mono Basin and a major administrative reallocation that could result from a court-mandated review of water allocations in the Sacramento/San Joaquin delta.

Virtually all segments of the California water industry concede that water transfers are a desirable way to reallocate water, and the state legislature has passed laws intended to facilitate such transfers. Despite this fact, transfers have not occurred on a major scale and have made only marginal contributions to the solution of the state's water problems. In fact, there is less transfer activity in California than in most other western states. The reasons behind this hesitancy are primarily institutional—existing water institutions tend to protect existing uses. Transfers of riparian and appropriative water rights held by individuals are constrained because many of these rights are tied to land or are unrecorded (or only partially recorded) as to quantity. Large blocks of appropriative rights are held by the Bureau of Reclamation and the California Department of Water Resources and allocated under contract to cooperatives of water users. The transfer of these contractual entitlements has been discouraged, in part because these agencies have not had strong affirmative policies supporting transfer and reallocation.

The southern portion of California's Central Valley, the San Joaquin Valley, is a particularly important and relevant target for transfer activity. The vast majority of developed water supplies in the valley are used to support agriculture. Urban water users in southern California recognize that some of the valley's agricultural supply represents a potentially inexpensive source of water to augment domestic supplies in the face of rapid population growth. At the same time, persistent ground water overdraft in some parts of the valley and sharp population increases projected for the valley's urban areas suggest that the valley economy itself will require additional water to maintain existing levels of economic activity and to support growth.

THE SETTING

California's great Central Valley occupies much of the northern and central heartland of the state. As shown in Figure 10.1, the valley extends more than 400 mi (640 km) from Redding in the north to Bakersfield in the south. The valley is bordered by the Coast Range and the Sierra Nevada mountains on the west and east, respectively; by the Klamath and Trinity mountains in the north; and by the Tehachapi mountains in the south. The valley's climate is characterized by dry, hot summers and mild, rainy winters. Average annual precipitation ranges from 38 in. (92.7 cm) in the north to 6 in. (14.6 cm) in the south. The alluvial soils have helped to make the valley one of the most productive agricultural regions in the world.

The valley is divided into three hydrologic basins, each of which drains roughly one-third of the area. The northern third drains to the Sacramento River, and the middle third drains to the San Joaquin River. The southern third, the Tulare Lake hydrologic basin, flowed to the San Joaquin River centuries ago but is today a closed basin

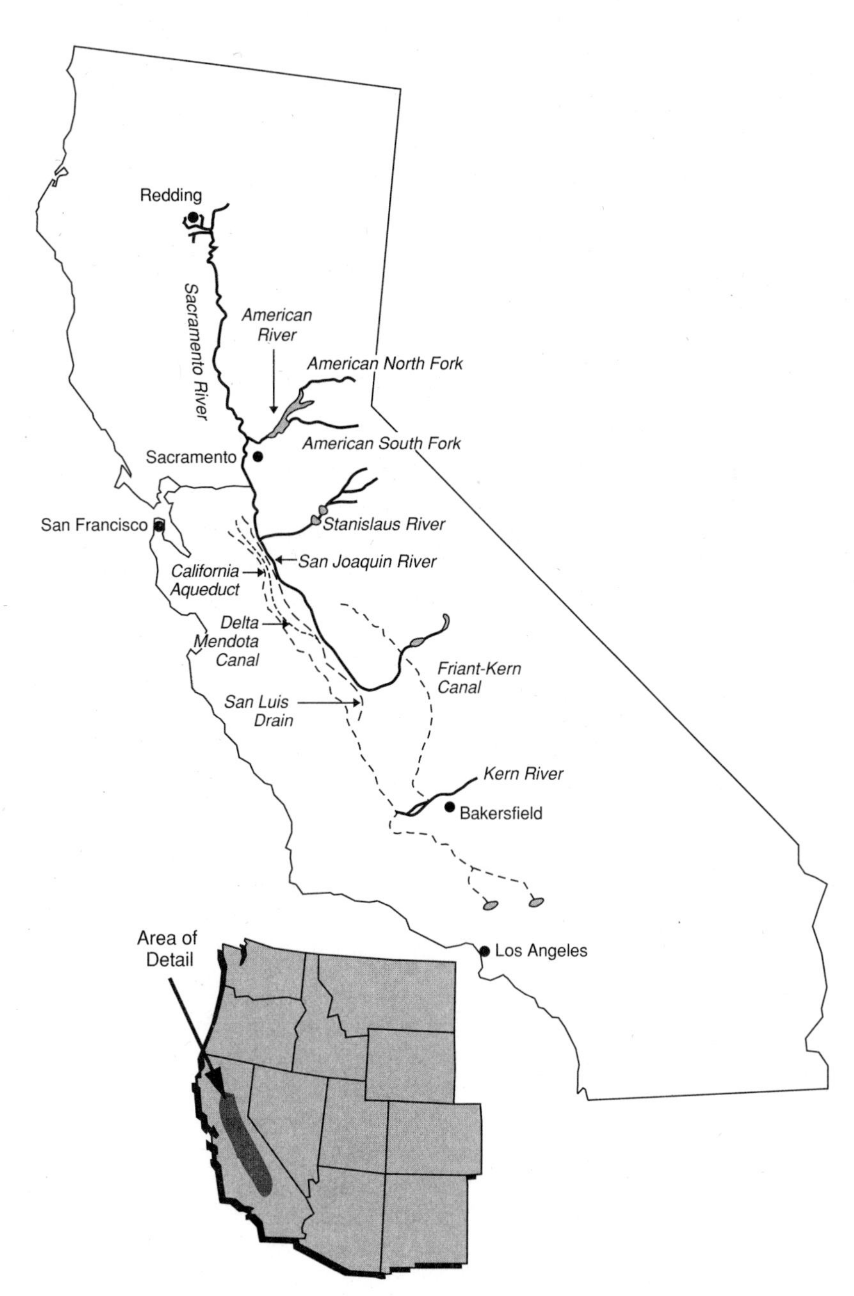

FIGURE 10.1 Main waterways and features, Central Valley of California.

except in extremely wet years. This case study focuses on the San Joaquin and Tulare Lake basins, which are frequently combined and characterized as the San Joaquin Valley. The Sacramento Valley poses some different and more localized transfer issues, which have been discussed elsewhere (Gray, 1990).

Irrigated agriculture is the dominant economic activity in the eight counties of the San Joaquin Valley. Approximately 4.7 million acres of land in these counties has been developed for irrigation (California Department of Water Resources, 1987b). In 1987 the value of agricultural production in the San Joaquin Valley totaled approximately $9 billion. The leading agricultural commodities include cotton, grapes, citrus, almonds, stone fruit, and dairy and livestock, each valued well in excess of $100 million in 1987 (American Farmland Trust, 1989).

In an average year, approximately 14.5 million acre-feet (17.9 million ML) of water is consumed to support irrigated agriculture in the San Joaquin Valley (California Department of Water Resources, 1987b). This is a little more than half of the water used by agriculture statewide and nearly 41 percent of the state's total net water use (California Department of Water Resources, 1987b). The valley's water comes from four sources: (1) locally developed surface supplies, (2) ground water, (3) the federal Central Valley Project (CVP), and (4) the State Water Project (SWP).

Local surface supplies, which account for approximately 36 percent of the net use in the San Joaquin Valley, arise predominantly in the Sierra Nevada. In the early years of irrigation, settlers built simple facilities that permitted them to divert water directly from streams to their fields. Subsequently, the substantial seasonal and annual variability of streamflows led irrigation districts to build sophisticated facilities to capture wet season and wet year flows for use during dry seasons and dry years.

The San Joaquin Valley overlies a substantial ground water resource, and ground water extractions account for about 20 percent of the valley's net water use. Overdrafting is a persistent problem, and the California Department of Water Resources (1987a) estimates that it amounts to 1.4 million acre-feet (1.7 million ML) in an average year and perhaps four times that amount in periods of severe drought. Despite the unattenuated overdrafting, many irrigation districts in the central and eastern portions of the valley have devised effective conjunctive use programs in which wet season and wet year flows are stored in underlying aquifers for use during dry periods.

Imported surface water supplies, which account for about 22 percent of net water use, were first made available through the Central

Valley Project, built by the Bureau of Reclamation during the 1940s and 1950s. The southern portion of the CVP, which is relevant here, includes two divisions. The Friant Division's dominant features are the Friant Dam on the San Joaquin River and the Friant-Kern Canal, which runs 150 mi (240 km) south to a point near Bakersfield. This division delivers an average of 1.5 million acre-feet (1.8 million ML) annually to approximately 15,000 farms on the east side of the San Joaquin Valley. The San Luis Division includes the San Luis Reservoir and a portion of the California Aqueduct, both of which are jointly used by the CVP and the state of California. This division services an area on the western side of the San Joaquin Valley, delivering approximately 1.2 million acre-feet (1.4 million ML) annually.

Surface waters are also imported to the San Joaquin Valley from the Sacramento basin by the State Water Project; this water is delivered to the western side of the San Joaquin Valley from the delta of the Sacramento and San Joaquin rivers via the California Aqueduct. The aqueduct, which lies on the western side of the valley, ultimately traverses the Tehachapi Mountains and enters the Los Angeles basin.

The water supply facilities of the San Joaquin Valley, as well as those of the state as a whole, are interconnected in a variety of ways. The Cross Valley Canal connects the California Aqueduct with the Friant-Kern Canal and allows water to be delivered from the Sacramento-San Joaquin delta to users in the Friant Division. The Kern River-California Aqueduct Intertie allows water flowing in the lower reaches of the Kern River to be put into the California Aqueduct and made available to users along the aqueduct. Ground water recharge basins are operated throughout the eastern and southern portions of the San Joaquin Valley and permit surface and ground water supplies to be managed conjunctively. These facilities, together with numerous smaller, more localized interconnections, provide a physical infrastructure that allows the water delivery systems of the valley to be managed in a highly integrated fashion.

The existence of physically integrated water delivery systems and the ability of different irrigation districts to exchange water from different sources mean that water can be moved between virtually any two points in the valley. In addition, water can be obtained from or transferred to most locations in the state with the exception of the North Coast basin and some central coast areas. Thus, except in very rare cases, water transfers in California need not be constrained by the absence of transport facilities. This contrasts with circumstances in other states, such as Arizona, where the potential for water transfers is crucially conditioned by the availability of conveyance facilities.

WATER INSTITUTIONS

California has a bewildering array of water institutions that shape the allocation of water and affect the capacity of water users to transfer, exchange, or lease water and water rights. The net effect is that water is allocated in large blocks controlled by strong local districts. The institutions that most critically affect the allocation of water are the water agencies and the system of water rights.

Strong federal, state, and local agencies influence the water agenda for the state. The primary federal agency is the Bureau of Reclamation, which operates the Central Valley Project. The major state agencies are the California Department of Water Resources, which is primarily responsible for water supply facilities in California, and the State Water Resources Control Board, a regulatory agency responsible for the administration of water rights and water quality programs. Because of the state's historical role as a primary provider, the regulatory role is less well developed than in other states.

California is unique in that much of its irrigable land, including more than 90 percent of such land in the San Joaquin Valley, falls within the boundaries of special purpose agencies—mostly water and irrigation districts—that have been formed to acquire and purvey water to local users. Statutorily, these districts function as special-purpose local governments, having been established under a variety of provisions in the California Water Code for the general purpose of acquiring, storing, and conserving water. The districts function much like "user cooperatives." Some have developed their own surface water supplies, whereas others contract for surface supplies with either the state of California or the Bureau of Reclamation. It is not uncommon for districts to have a combination of locally developed supplies and supplies delivered under contract with either the state or the federal government. These districts influence the terms under which virtually all of the irrigation water in the valley is used. Their financial and political powers give them key roles in facilitating or inhibiting transfers.

California's system of water rights also has a major effect on the extent to which transfers occur. In no other part of the West has so much development taken place in the face of so much theoretical confusion and uncertainty about entitlements. The legal system has survived because of the endless promise of aqueducts from distant places—the mountains, the Colorado River, the North Coast, the Columbia River. The state is unique in its dual system of water rights, in which both riparian and appropriative rights exist side by side, largely unintegrated with ground water rights (Governor's Commission to Review California Water Rights Law, 1978; Hutchins, 1956). Riparian rights to surface water are attached to any tract of land adjoining a stream or lake. A riparian right is limited to the quantity of water that can be put to "reasonable and beneficial" use on the land in question. The right cannot be expanded by adding dry land to the tract but can be diminished if the tract is subdivided and the size of the parcel that abuts the stream or lake is reduced. Where the sum of riparian rights exceeds the common water supply, rights holders are obliged to share the water equitably. Riparian rights are held almost exclusively by private individuals, because public agencies and private companies cannot acquire them except for use on lands that they own.

Two features of riparian rights are especially significant. First, in virtually all instances, riparian rights take precedence over appropriative rights. That is, a typical appropriative right can be exercised only on water that is surplus to water subject to riparian rights. In time of shortfall, riparian rights are superior to all appropriative rights except for the very few that were established prior to the riparian rights (Bowden et al., 1982). However, the California Supreme Court has held that unexercised riparian rights may be subordinated to appropriative rights in an adjudication. Second, riparian rights typically may not be transferred independently of the land to which they attach. The water may be captured through adverse possession, but in such instances the riparian right is not transferred; instead, a prescriptive right, similar to an appropriative right, is established in its place. It is thus clear that under the prevailing legal doctrine in California a riparian right cannot be traded except in association with the land.

Appropriative rights to surface waters are obtained by using the water continuously and for a reasonable, beneficial purpose. Failure to use the water for a period of 5 years results in loss of the right. Appropriative rights can be established only for water that is "surplus" to the reasonable and beneficial requirement for both riparian and senior appropriative rights established in the water source. Seniority in appropriative rights is established by the first date of appropriation, with an early right being senior to a later one. Today, appropriative rights are licensed by filing with the State Water Resources Control Board, which then holds hearings and presents findings with respect to the availability of surplus water, the absence of injury to third parties with standing, and the "reasonableness" of the appropriation. The licensing process has been in effect only since 1914, and rights established prior to that date have not been recorded. Thus the ownership and quantity of such rights are subject to dispute, making them potentially difficult to transfer.

Appropriative rights may be established by private users or by intermediaries such as local public agencies or private water companies. In the latter case, the members of the intermediary are viewed as having a beneficial interest in the water. Thus appropriative rights to surface water tend to reside ultimately in the users of the water (Bain et al., 1966). The appropriative right is not inherently tied to the land, and the site of use or point of diversion may be changed subject to the approval of the State Water Resources Control Board.

However, there is a tendency for appropriative rights to become in fact appurtenant to the land on which they are used, which is specified on the license issued by the state. Although it may be a simple matter to change the kind of use—from irrigation to municipal and industrial use, for example—changing the location of use is more problematic. The law recognizes the interrelatedness of uses, and changes in the site of use or point of diversion cannot occur without approval from the State Water Resources Control Board. This approval is not granted if unreasonable injury will result to third party users.

Ground water is also subject to a separate system of water rights. The correlative right entitles holders to use ground water on overlying lands and requires that shortfalls be shared equitably. Correlative rights cannot be established for lands not overlying the aquifer. For nonoverlying lands, appropriative rights to ground water—similar in most respects to appropriative rights to surface water—may be established. Appropriative rights to ground water are inferior to correlative rights.

A critical difference between ground and surface water rights is that there is no requirement for filing and licensing to establish a ground water right; it is necessary only to initiate use and ensure that it is continuous. The result is that ground water rights are not recorded or quantified except in a few urban basins in southern California where there has been extensive litigation. In the San Joaquin Valley, as in most other agricultural areas of California, the permissiveness of ground water law has fostered a situation in which there are virtually no restrictions on ground water pumping other than the economic restrictions imposed by cost.

Permits for a significant portion of the appropriated surface water in the San Joaquin Valley have been acquired from the State Water Resources Control Board by the Bureau of Reclamation (for the Central Valley Project) and the Department of Water Resources (for the State Water Project). This water is not used directly by the appropriating agencies but is almost entirely sold under contract to water districts and water agencies. As a consequence, water users in the San Joaquin Valley who receive water from the Department of Water Resources or the Bureau of Reclamation hold contractual rights with those agencies rather than water rights per se. Contract holders clearly have equitable entitlements to water deliveries, but the existence of a contractual relationship with a government entity obligated to repay the costs of the project complicates attempted transfers.

The permits for the Central Valley Project define the place of use as the entire service area of the CVP and allow the water to be used for many purposes (Kahrl, 1979). Under the terms of the permit, one CVP contractor can transfer water to another CVP contractor within a service area without securing the approval of the State Water Resources Control Board to change the place of use, the purpose of use, or the point of diversion. As a matter of bureau policy, such transfers are generally limited to a one-year period (Gray, 1990).

These rules provide water users in individual service areas with flexibility to respond to changing water conditions. Transfers between CVP contractors within the same division are quite common. However, current bureau policies (somewhat modified by the 1991 drought) do not encourage transfers between divisions. By limiting permissible transfers to intradivisional exchanges among contractors, the bureau, in effect, restricts exchanges that involve changes of use because virtually all CVP users are engaged in irrigation. Transfers of CVP water to users outside the CVP service area have not occurred so far. This limitation, apparently based largely on concerns about potential third party effects, creates a substantial barrier to transfers between agricultural and urban users. No CVP water has yet been transferred to municipal and industrial users in the southern California urban areas, nor has any been transferred to urban users in the San Francisco Bay area who are not themselves bureau customers.

The State Water Project contractors have not established a routine water transfer system. Although there are several long-term (50 years) exchange arrangements that permit districts remote from the SWP aqueduct to receive other water in exchange for their SWP entitlement, there have been virtually no other transfers, short or long term, among SWP contractors (Gray, 1990). This is attributable to the considerable uncertainty surrounding the rules, processes, and policies governing water transfers.

Ironically, California has perhaps the most supportive water laws related to transfers. These laws, most of which were enacted during the 1980s, manifest the legislature's conclusion that water reallocation through private transactions is one significant, low-cost means by which the state can respond to its intensifying water scarcity. Yet despite the presence of detailed water transfer law, California has had less water transfer activity than most of the other states examined by this committee. Only 24 petitions to transfer water were submitted to the State Water Resources Control Board between 1981 and 1989 (Gray, 1990). Of these, 19 were approved. By contrast, during the same period, approximately 1,200 transfers between CVP contractors occurred. None of these transfers required the approval of the board, and virtually all were for periods of less than a year. Transfers among CVP users appear to be an accepted means of dealing with variations in the demands of individual irrigators from year to year. It is important to recognize, however, that all of these exchanges were limited to transfers of water—there were no transfers of water rights (MacDonnell, 1990).

The fact that transfer activity in California seems confined to short-term exchanges within project service areas cannot be attributed to the lack of profitable opportunities for trade. Several studies demonstrate that significant opportunities for profitable water transfers between the agricultural regions and urban areas exist. For example, Vaux and Howitt (1984) showed that in 1980 some 2 million acre-feet (2.5 million ML) could have been profitably traded from agricultural regions to the state's major urban areas. These quantities were projected to grow to more than 2.5 million acre-feet (3.1 million ML) by 1995. This study showed that buyers and sellers would have jointly benefited by $66 million in 1980, with benefits growing to $156 million by 1995. Other studies document the existence of profitable trading opportunities within agricultural regions such as the San Joaquin Valley (California Assembly Office of Research, 1985; Vaux, 1986).

The explanation for the lack of transfer activity appears to lie instead with water institutions and a reluctance to change those institutions. Although a number of factors can inhibit transfers, three stand out. First, the lack of a coherent transfer policy for the CVP and legal uncertainties associated with policies governing transfers of SWP water have dampened transfers between the service areas of these projects and effectively prohibited transfers to users outside the institutional and geographic realms of the projects. These restrictions and uncertainties are grounded in concerns about the need to protect existing uses of water, possible adverse third party effects, and resistance to change in water management and allocative institutions.

Second, the profitability of many potential trades is eroded by the expectation of high transaction costs. As noted earlier, many appropriative rights in California have not been clearly quantified, and costly adjudications may be prerequisite to exchange of these rights or waters. Moreover, there is so much uncertainty surrounding the process through which transfers can be consummated—uncertainty over who has standing to intervene and uncertainty over when interveners are entitled to block a transfer and when compensation should be paid—that the transfer process suffers. Much of this uncertainty is rooted in the lack of clear rules and procedures for dealing with third party effects.

Third, perhaps the most pervasive explanation for the lack of transfer activity is the failure of water institutions to evolve in ways that facilitate transfers. California's water institutions traditionally have dealt with water scarcity through the construction of dams and canals. Historically, the Department of Water Resources and the Bureau of Reclamation have performed an entrepreneurial role in developing and marketing water supplies. Neither agency has ever attempted to promote and facilitate a broadly based system of water transfers or administrative reallocation.

The State Water Resources Control Board, although possessing broad authority to restrict and reallocate both surface and ground water, has not exercised that authority in the way that state engineers in other western states have typically exercised it (Gray, 1989). In the rare instances where the board has addressed allocative issues (see Chapter 11), it has relied on the threat of using the authority rather than wielding it directly. (The efforts of the board to reallocate water in the Sacramento/San Joaquin delta, in response to a court mandate, could be the beginning of a substantial departure from its historical reluctance to be involved in water reallocation issues.)

The fact that California's State Water Project remains uncompleted means that many of its contracting districts receive less water than the formal entitlements established by the contracts. Moreover, in the absence of additional water project construction, the difference between what the districts actually receive and the contracted quantities will grow. These contractors are shareholders in a pool of water, and as long as there is more demand than the pool can supply, the contractors will resist any effort to transfer water (Miller, 1990). It is not clear whether claims based on this argument are legally enforceable (Gray et al., 1990), but the consequent uncertainty creates a substantial impediment to transfers outside the SWP service area. The failure to resolve the third party standing of other state contractors means that any external trade could be subject to costly litigation, and few buyers are willing to risk incurring those costs.

Local water districts have both the capability and the incentive to preserve the historical allocation of water through a variety of legislative and legal means. This contributes to the persistence of institutions that resist efforts to deal with water scarcity through reallocation.

Given current institutions and the absence of clear procedures governing water transfers, transfers can be accomplished now only when there is complete consensus that a transfer is beneficial to some parties and harmless to all other parties. The emergence of the consensus process stems from the widely held perception that any party to a transfer, no matter how remote, can effectively veto the transfer by demonstrating some adverse impact. To exercise such a veto, an individual or group need not resort directly to courts or administrative agencies. Rather, the mere threat of prolonged and costly litigation is usually enough to prevent agreements. Although the "complete consensus" process may work in the short run for a few trades, it is not a recipe for success in the long run. Apart from the fact that change is rarely without cost, there is the larger issue of self-interest. Paralysis can result when a polity dominated by special interest groups seeks to redistribute or reallocate wealth (Thurow, 1980). In short, it is in the self-interest of any group that is adversely affected by a public policy or action to resist and block that action. Thus, with the exception of a few transfers that are broadly profitable, a consensus process in which there are no mechanisms for resolving disputes is a major institutional impediment to water trading.

Existing institutional arrangements are not immutable, but they are not likely to cede the power associated with the control of large blocks of water without considerable external pressure. However, the fact that California's major water agencies have not actively encouraged transfers as one means of alleviating water scarcity has dampened the impetus for institutional change. The failure of all but a few political leaders to advocate water transfers explains, at least in part, the absence of a proactive approach to transfers by the major water agencies.

Representatives of the Bureau of Reclamation indicate that the U.S. Department of the Interior's 1988 transfer policy has placed the bureau in a largely passive position. The California Department of Water Resources has not actively supported long-term transfers, perhaps out of concern about becoming embroiled in legal conflicts. And the major urban water agencies, which are potentially the largest buyers of agricultural water supplies, also have been quite cautious in their approach to transfers. This is due, at least in part, to fears of resurrecting the "Owens Valley syndrome"—the perception that large urban water agencies are "stealing" water from rural citizens.

In the absence of political leadership, the institutional changes required to facilitate water transfer are unlikely to happen soon. The major water purveyors in the state appear reluctant to encourage institutional change without such leadership. Thus, as current institutional arrangements persist, it will be difficult for California to realize the full potential of water transfers as a strategy to cope with water scarcity.

THIRD PARTY IMPACTS

In a state as vast and diverse as California, examples of almost every type of third party effect can be identified from among the array of potential transfers. Some of these effects are more important and pervasive than others, however. For example, adverse cultural impacts on minority groups and effects arising from undefined or poorly defined Indian water rights are likely to be far less significant in the San Joaquin Valley than elsewhere in the West. Issues relating to the impact of water transfers on minorities exist, but the impacts are not as well identified as they are in places such as New Mexico and the case for giving them substantial weight is not as great. By contrast, the two most prominent classes of third party effects involve areas of origin and the environment.

Areas of Origin

Several impacts on areas of origin are illustrated in the case involving the Berrenda Mesa Water District and its parent institution, the Kern County Water Agency. The Berrenda Mesa Water District lies in Kern County in the southwestern region of the San Joaquin Valley, and it depends on the SWP for its entire water supply. The Kern County Water Agency is a distinct type of special district that overlies all of Kern County but whose member districts, including Berrenda Mesa, are only those with entitlements to state water. The Kern County Water Agency was formed for the purpose of contracting with the state of California for water from the SWP and has access to the entire tax base of Kern County in financing its operations. Its primary function is to sell water from the SWP to member districts. Together, the individual member districts and the Kern County Water Agency determine the allocation of all waters imported to Kern County via the SWP (Vaux, 1986).

In 1986 the Berrenda Mesa Water District foreclosed on the owners of approximately 6,500 acres of land because of their continuing inability to meet the payments required by the water service agreements with the district. In the aftermath of the foreclosure, Berrenda

Mesa's remaining water users were then faced with the possibility of having to assume the portion of the district's fixed costs previously borne by the foreclosed users. Berrenda Mesa estimated that operations covering an additional 6,000 acres might be financially threatened if this occurred. The district thus attempted to market the water released via the foreclosure and thus help to defray the fixed costs.

The Kern County Water Agency and other state water contractors opposed the sale of Berrenda Mesa's water, arguing that they were entitled to any water that was surplus to the demands of an existing contractor. The argument of the Kern County Water Agency differed in some important respects from arguments mounted by other water districts. The Kern County Water Agency assesses all taxpayers in Kern County about 15 percent of the cost of delivering water to districts where agricultural uses predominate. The rationale for this assessment is based on the proposition that most agricultural lands will be irrigated in perpetuity and thus current landowners should not be obligated to pay the full cost of the water. Rather, all citizens of Kern County have an interest in ensuring that water currently allocated to Kern County remains there, both to preserve the important economic base provided by irrigated agriculture and to support further urban growth in Kern County.

This case illustrates at least three types of third party effects, two of which are related to areas of origin. First, when a water user (or users) within a district proposes to transfer water to users outside the district, real financial costs can accrue to other users within the district if nothing is done to defray the capital costs of facilities and other fixed costs previously borne by the transferor. In the extreme, this increased burden could harm the economic viability of district operations and, perhaps, bankrupt the entire district. From an economic perspective, it can be argued that each grower who joins a district or collective binds himself, to some extent, to the welfare of the collective. If the collective as a whole is not economically viable, efficiency considerations dictate that it should fail unless some subset of the collective is profitable enough to defray all fixed costs. Although the efficiency implications of such situations seem quite straightforward, there are substantial issues of equity.

The crucial issues turn on the circumstances under which the individual is bound to the collective. Normally, such questions would be addressed in the collective's charter. However, most water districts in the West were organized before water marketing was viewed as a practical option, and their charters may thus be silent on the rights of members to transfer water to remote buyers. (It should be noted that the Department of Water Resources would be unlikely to approve any trade that threatens the financial viability of the SWP itself, but this protection would not necessarily extend to the financial viability of individual districts, particularly when a buyer is willing to defray the selling district's contractual obligations to the state.)

Some western water districts (e.g., the Northern Colorado Water Conservancy District) prohibit members from selling or leasing water to buyers outside the district. This type of prohibition protects the interests of district members but may create significant allocative inefficiencies by barring transfers that are otherwise profitable. When sellers insist that any external buyer retire the costs of conveyance capacity or facilities idled by the transfer, allocative distortions are introduced, because buyers are required to defray the costs of the seller. That is, if the buyer is required to pay for some of the prior financial obligations of the seller, which are unrelated to the value of the water, less water would be transferred than would be economically optimal.

A second concern related to areas of origin is the indirect impact of water transfers on firms and industries linked to irrigated agriculture and the impacts on agricultural communities. For example, in an area where agricultural production is reduced because of transfers of water to other regions, the remaining production may prove insufficient to support some or all of the local packinghouses or seed, fertilizer, and machinery distributors. Similarly, as irrigated agriculture declines, the community becomes less prosperous, and its economic and social infrastructure may decline. Banks, pharmacies, and other essential firms close. The social structure provided by churches, civic groups, and political organizations weakens just when community members most need such support.

Area-of-origin concerns often focus on the multiplier effects of disinvestment in irrigated agriculture. Just as investments in irrigated agriculture bring "linked industries" to an area, disinvestment causes them to leave. The disinvestment, then, leads to other adverse consequences for the social and political structure of the community as the tax base shrinks, community services diminish, and unemployment rises. The prospect of this chain of impacts raises fears of impoverishment and fears that traditional agricultural lifestyles may vanish.

The Kern County Water Agency was in a unique position in regard to this latter type of concern because taxes that it had collected were used to defray about 15 percent of the costs of the water that Berrenda Mesa wished to sell. Therefore the agency could argue that it had recognized indirect benefits stemming from irrigated agriculture in Berrenda Mesa and had borne a portion of the water supply costs as a consequence. (The question of whether the level of the Kern County Water Agency's financial contribution appropriately reflected the magnitude of the indirect benefits is a separate issue.) In most cases, local (or regional) taxpayers do not participate in the financing of water supplies, although some or all of them may reap indirect benefits from irrigated agriculture. Nevertheless, in many instances, local firms and communities would suffer both economic and social costs as a consequence of water transfers.

A third issue raised by the Berrenda Mesa/Kern County Water Agency example relates to the lack of clear rules and procedures governing the transfer of water. The Berrenda Mesa water sale was opposed by other SWP contractors who argued that they had first call on any surplus SWP supplies. The contractors argued, in effect, that the terms of SWP contracts would make them third parties to any transfer of water outside the SWP service area. This position was never tested in the courts, presumably because potential buyers feared the high costs of litigation. However, if the contractors had prevailed in such a court test, Berrenda Mesa could have lost the water and received no compensation.

Ultimately, Berrenda Mesa was unable to sell its water because the terms under which water from the SWP could be sold or exchanged were never clearly specified and the district was unwilling to risk a test case in the courts. Subsequently, the Kern County Water Agency implemented a procedure under which it guarantees a sale price for the surplus water to Berrenda Mesa during hydrologically "normal" years. Berrenda Mesa uses the water in "dry" years and is essentially stuck with it in hydrologically "wet" years. Thus Berrenda Mesa has been effectively barred from selling at least a portion of its surplus water by the actions of the Kern County Water Agency and the threats of other SWP contractors.

This case emphasizes how the lack of clear, legally tested rules and procedures can inhibit water trading and illustrates the need to specify precisely, perhaps as a matter of law, the groups and individuals who are entitled to third party standing in any water transfer. Kern County residents aside, there is no a priori reason why other state water contractors should be accorded third party status in this situation.

Environment

Water transfers in the Central Valley could have a broad range of environmental impacts. On the one hand, some environmental groups have supported transfers as a means of avoiding the adverse environmental impacts associated with the construction of new storage facilities. On the other hand, an unfettered market-like system of water transfers could bring environmental damages. Unless transfers are restricted to water that is used consumptively or otherwise becomes unusable, runoff that flows back to watercourses could be diminished, and instream flow problems would follow. Downstream rights holders might also be harmed.

In California, rights to instream flows are not vested in any specific individual, group, or state agency. Unlike many other states, there are no exclusive property rights to instream flows. They are simply common property rights, which may be diminished by private uses. Although groups wishing to purchase water to augment instream flows could, in theory, acquire water for those purposes, it is not clear how they could protect that water from downstream appropriation. Since most instream uses of water are public goods, it is fair to conclude that a market system without some form of public intervention would underallocate water for instream flows.

Instream flow protection has always been ad hoc in California. During the last decade the California Supreme Court has employed the public trust doctrine to protect environmental uses (see, for example, *National Audubon Society* v. *Superior Court,* 1983). Just as the doctrine was used to limit appropriations in Mono Lake basin for export to Los Angeles, it could be invoked either administratively or through the courts to bar or modify proposed transfers from one basin to another to maintain desired flows within the basin of origin or current use. The legal procedures involved would be both costly and time-consuming, and the threat of such proceedings might be sufficient to block transfers, even those with relatively benign environmental effects. The fact that the implications of the public trust doctrine have not been fully elaborated by either the courts or the State Water Resources Control Board contributes to the uncertainty surrounding proposed water transfers. If transfers are to be facilitated, clear rules and procedures governing environmental impacts will need to be developed.

Such rules and procedures should recognize explicitly that water transfers in certain local situations may represent both an efficient and an equitable way of resolving an environmental problem. The appearance of toxic concentrations of selenium in drainage waters pooled at Kesterson Reservoir in the northwestern San Joaquin Valley has been carefully documented elsewhere (National Research Council, 1989). That situation illustrates how irrigation drainage waters can degrade ground and surface water quality and harm fish and wildlife. Part of the solution to problems posed by drainage waters involves more careful management of irrigation water to reduce the volume of drainage produced (Letey et al., 1986). The existence of well-functioning water markets could help facilitate the management of drainage waters.

Overirrigation in a number of water districts on the west side of the San Joaquin Valley contributes to the drainage problem (University of California Committee of Consultants on Drainage Water Reduction, 1988). This overirrigation can be explained in part by the lack of incentives to manage irrigation water more intensively and in part by the inability of some growers to finance the capital and labor necessary for more intensive management. The Reclamation Reform Act of 1982 requires all districts receiving water from the Bureau of Reclamation to develop water conservation plans but makes no provision for recovering the costs expended. The establishment of water markets might help address both problems, first, by providing an incentive to sell or lease water saved as a consequence of more intensive management and, second, by providing a source of revenue to finance water-saving technology and management. There may also be opportunities for buyers to participate directly in the financing of drainage-reducing measures in a fashion similar to those discussed in the Imperial Valley case study.

Although transfers can facilitate reductions in the volume of drainage waters, they do not offer a complete solution to the drainage problem. Means will still have to be found to maintain salt balances by either exporting salt from the basin or sequestering it in some harmless way. By generating funds to finance salt management activities, transfers could be helpful in supporting solutions to the problem of where to dispose of salts.

CONCLUSIONS

1. The lack of clear rules and procedures pervades the water transfer process in California. Although the legislature has passed a number of laws designed to facilitate water transfers and the Department of Water Resources, particularly in its recent creation of a drought-induced water bank, has made efforts to implement those laws, legal and procedural uncertainties constrain transfers. Uncertainties with respect to possible third party effects are particularly significant, since it is unclear which third parties would have to be accommodated in connection with transfers. For example, the ability of State Water Project contractors to veto transfers to noncontractors has not been resolved completely and the resulting possibilities for litigation will

impede trade. The California situation illustrates the extent to which failure to distinguish between those third party impacts that must be accounted for in a transfer and those that may safely be ignored can impede water reallocation through transfers.

2. Institutions with a stake in developing new water supplies tend to resist efforts to reallocate water, which also impedes development of the transfer process. Water users and their cooperatives can be especially effective in blocking transfers by identifying all manner of potentially adverse third party impacts. The lack of clear policy governing which third party impacts should be considered means that all such impacts tend to be given weight in a transfer proceeding, irrespective of their merits.

Under California water law, riparian rights and pre-1914 appropriative rights are particularly difficult to transfer either because they are appurtenant to the land or because they have not been quantified. The apparent restrictions placed on transfers by the Bureau of Reclamation and the lack of clear policy governing external exchanges by SWP contractors also tend to restrict transfers of contract rights or entitlements outside or even within their respective project service areas.

3. Several types of potentially adverse impacts on areas of origin are illustrated by the Berrenda Mesa case. To what extent should the financial viability of an irrigation district be protected in the transfer process? To what extent should local economies be protected? On efficiency grounds alone the answers are relatively clear. The equity impacts are more difficult to assess but will clearly be important.

Areas-of-origin effects are probably dealt with most effectively through broad state policies that articulate the extent to which water-dependent rural economies should be preserved. Such policies should be developed with a clear understanding of the costs involved. Thus, for example, a policy that tends to impede the reallocation of water from areas of origin may require the construction of new facilities to supply the water necessary to support new growth. In formulating such policies, it will also be important to recognize that disinvestment in unprofitable industries is one of the ways by which mixed free market economies maintain their vitality. If the economy of the West is to remain vibrant, and if national and world demands for food and fiber produced in the West do not grow substantially in coming decades, disinvestment in irrigated agriculture is probably inevitable and efforts to forestall it are likely to be counterproductive in the long run.

Similarly, political judgments will have to be made about equity effects that arise when a sale by several water users in a cooperative imposes increased costs of indebtedness on the remaining water users. From a strict efficiency standpoint, there is no reason why the financial integrity of such a cooperative should be protected. However, issues of equity and financial integrity are real and must be addressed. They are probably best addressed by the user cooperatives themselves.

4. The environmental impacts of water transfers are varied. Policies that give standing to instream uses and provide for the public purchase and protection of water for environmental uses are needed. The fact that water transfers provide a means for alleviating some environmental problems, as illustrated by the potential of market-like transfers to help resolve the San Joaquin Valley drainage problem, suggests that policies should be crafted to provide balance between environmental and consumptive uses.

REFERENCES

American Farmland Trust. 1989. Eroding Choices—Emerging Issues: The Condition of California's Agricultural Land Resources. San Francisco, Calif.

Bain, J. S., R. E. Caves, and J. Margolis. 1966. Northern California's Water Industry. Baltimore, Md.: The Johns Hopkins Press.

Bowden, G. D., S. W. Edmunds, and N. C. Hundley. 1982. Institutions: Customs, laws and organizations. In Ernest A. Engelbert and Ann Foley Scheuring, eds., Competition for California Water. Berkeley: University of California Press.

California Assembly Office of Research. 1985. Water Trading: Free Market Benefits for Exporters and Importers. Report 058-A. Sacramento.

California Department of Water Resources. 1987a. California Water: Looking to the Future. Bulletin 160-87. Sacramento.

California Department of Water Resources. 1987b. California Water: Looking to the Future. Statistical Appendix. Bulletin 160-87. Sacramento.

Governor's Commission to Review California Water Rights Law. 1978. Final Report. Sacramento, Calif.

Gray, B. E. 1989. A primer on California water transfer law. Arizona Law Review 31:745.

Gray, B. E. 1990. Water transfers in California: 1981-1990. Unpublished report. Hastings College of Law, San Francisco, Calif.

Gray, B. E., B. C. Driver, and R. Wahl. 1990. The transferability of water provided by the State Water Project and the Central Valley Project: A report to the San Joaquin Valley Drainage Program. Unpublished manuscript. Hastings College of Law, San Francisco, Calif.

Hutchins, W. A. 1956. The California Law of Water Rights. Sacramento Office of the California State Engineer.

Kahrl, W. L. 1979. The California Water Atlas. Sacramento Office of Planning and Research.

Letey, J., C. Roberts, M. Penbreth, and C. Vasek. 1986. An Agricultural Dilemma: Drainage Water and Toxic Disposal in the San Joaquin Valley. Special Publication 3319. Oakland: University of California, Division of Agriculture and Natural Resources.

MacDonnell, L. J. 1990. The Water Transfer Process as a Management Option for Meeting Changing Water Demands. Vol. 1. Report submitted to the U.S. Geological Survey.

Miller, B. J. 1990. Water transfers in California: Problems and solutions. Unpublished manuscript. B. J. Miller Associates, Berkeley, Calif.

National Audubon Society v. Superior Court, 658 P.2d 709 Cal. (1983).

National Research Council (NRC). 1989. Irrigation-Induced Water Quality Problems: What Can Be Learned From the San Joaquin Valley Experience? Washington, D.C.: National Academy Press.

Thurow, L. C. 1980. The Zero-Sum Society: Distribution and the Possibility for Economic Change. New York: Basic Books.

University of California Committee of Consultants on Drainage Water Reduction. 1988. Opportunities for Drainage Water Reduction. Riverside: University of California Water Resources Center, Salinity/Drainage Task Force.

Vaux, H. J., Jr. 1986. Water Scarcity and Gains from Trade in Kern County, California. Pp. 67-101 in Kenneth D. Frederick, ed., Scarce Water and Institutional Change. Washington, D.C.: Resources for the Future.

Vaux, H. J., Jr., and R. E. Howitt. 1984. Managing water scarcity: An evaluation of interregional transfers. Water Resources Research (20):785-792.

11

California's Imperial Valley: A "Win-Win" Transfer?

Success can offer as many lessons as failure, so it is useful to conclude this series of case studies with a discussion of California's Imperial Valley. In early 1989 the Imperial Irrigation District (IID) and the Metropolitan Water District (MWD) of Southern California signed a water conservation agreement. Two other irrigation districts, the Palo Verde Irrigation District and the Coachella Valley Water District, became part of the agreement in late 1989 (IID and MWD, 1989). In brief, MWD will pay for a program of water conservation for IID. In return, IID will reduce its call on the Colorado River by the amount conserved, and MWD will be entitled to divert this amount into its system at Parker Dam. Although MWD prefers not to characterize the agreement as a water transfer, it is otherwise almost universally viewed as the first major rural-to-urban transfer of irrigation water in California and will be a model for future transfers that try to accommodate urban demands and preservation of the state's productive agricultural economy. It is important to realize that the Imperial Irrigation District-Metropolitan Water District transfer was an "easy" case because no existing users were displaced and third party effects were minimal or indirect enough to be ignored. Still, it offers an important illustration of the potential of transfers.

THE SETTING

The Imperial Valley lies on the northern edge of the Sonoran Desert in southern California about 50 mi (80 km) west of the Colorado River (Figure 11.1). The area was originally called the Salton Sink and, for a while in the nineteenth century, the Valley of the Dead. It was renamed the Imperial Valley at the turn of the century by the irrigation pioneer and promoter George Chaffey as part of an effort to attract settlers to the harsh desert.

The valley is surrounded on the north, east, and west by mountains, and the entire valley floor is below sea level. The U.S.-Mexican

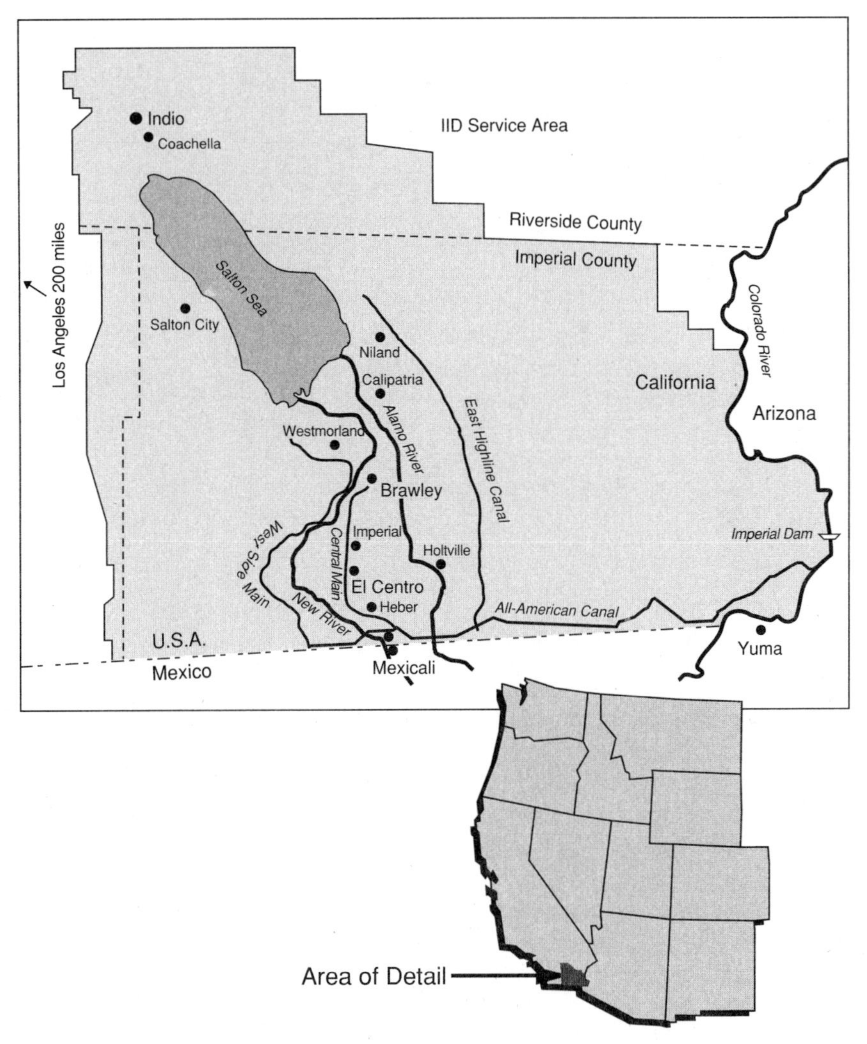

FIGURE 11.1 Main waterways and features, California's Imperial Valley.

border forms the southern boundary. Two rivers, the Alamo and the New, flow from Mexico into the Salton Sea at the western edge of the valley. These rivers are actually old shallow riverbeds into which the Colorado once flowed when its normal channel became silted up and it overflowed its banks in a southwesterly direction. It has been said that the Imperial Valley is one of nature's jokes—it contains hundreds of thousands of acres of flat fertile land formed by silt deposited in the ancient delta of the Colorado and the growing season is perpetual, but rainfall averages less than 3 in. (0.7 cm) per year. Ground water resources are minimal, so the valley, actually a geologic sink, must rely on the Colorado River for its supply.

Irrigation has been a great success in the valley. Approximately 497,000 irrigated acres produces a wide variety of crops. In 1988 some 349,281 acres was devoted to field crops, primarily alfalfa, cotton, Sudan grass hay, sugar beets, and wheat, with a gross value of $250 million. Much of the nation's winter melons and table vegetables come from the valley, and in 1988 some 119,682 acres was devoted to melons, lettuce, and other vegetables. These crops produced a gross value of nearly $460 million. Some land is devoted to citrus and dates. In addition, there is a large and valuable livestock industry, as well as significant honey production.

The lifeblood of the valley is the Colorado River, and the valley depends on its senior water rights to sustain itself. Because the basin is an inhospitable place, there was limited use of the river before the valley was promoted and settled between 1901 and 1904. Several Indian tribes, including the ancient Anasazi, practiced flood irrigation along the lower reaches of the river. Mormons and farmers in the Uncompahgre Valley of Colorado had made limited diversions starting in 1854, but California did not begin to use the river until Samuel Blythe developed the Palo Verde Valley upstream. Arizona developed the Yuma Valley in the 1880s, and Imperial diversions began in 1901 when the California Development Company posted a notice and diverted water that ran into a canal that mostly traveled through Mexico on its way to the valley.

Water allocation in California today is characterized by the storage and allocation of large blocks of water by federal, state, and local irrigation districts and other public entities. Much of this water is allocated to irrigated agriculture, but the continued growth of the state's urban areas and growing recognition of the need to incorporate instream flow protection into all decisions on water resource allocation create strong pressures for large-scale rural-to-urban transfers and the reduction of agricultural use. Despite its established water rights, its long ability to shape federal reclamation efforts to its benefit, and the high value of many of its crops, the Imperial Valley is now vulnerable to reallocation pressures. It is the last major agricultural area in southern California, its water use practices have long been questioned, and the value of crops grown in the valley varies greatly.

The history of the settlement and cultivation of the Imperial Valley is in large part the history of the regulation of the Colorado River and the growth of agribusiness in the state. As a 1991 appellate court opinion upholding the state's power to curb wasteful use practices in the valley observed, IID "has occupied a position of great strength, discretion and vested right in a geographical part of the country that is 'far western,' embracing a philosophy that is independent in every sense of the word. Recent trends in water-use philosophy and the administration of water law have severely undermined the positions of districts such as IID" (*Imperial Irrigation District* v. *State Water Resources Control Board,* 1991). That history has enabled the valley to chart its own destiny, but the valley is increasingly vulnerable to criticism that it uses a disproportionate share of southern California's water. Ninety-eight percent of Colorado River water deliveries to the IID go to irrigation, and the average use is more than 5 acre-feet (6 megaliters (ML)) per acre. A wide variety of irrigation techniques are used.

From the start the Colorado River was the basis of the area's survival. About 7,000 people settled in the valley between 1901 and 1904, but the great floods of 1905 and 1907 threatened the rapidly developing agricultural economy. The story of the valley's efforts to tame the river has been told many times, and historians are still debating its meaning (Hundley, 1975; Reisner, 1986; Worster, 1985). For the purposes of this case study, the important points are that the land development perfected some of the earliest rights along the Colorado River, but these rights could not be enjoyed without upstream flood control dams to protect the conveyance facilities from destruction by floods and sedimentation. In addition, the early canal from the Colorado ran through Mexico, which vigorously but unsuccessfully opposed the 1901 diversions. Mexican-Imperial tensions were the major source of many problems. Because the valley's water supply passed through Mexico, the valley faced problems such as pollution, increased financial burdens to maintain the canal, and the threat of revolutionary violence and disruption of supply.

The valley's fate was ultimately linked to the development of the entire Colorado River, although it resisted this linkage for two decades. Arthur Powell Davis, John Wesley Powell's nephew, proposed a series of large storage reservoirs along the river in 1902, when he joined the newly established federal Reclamation Service. Initially, most farmers wanted only an "all American" canal, a canal built to serve U.S. land, not a series of flood control and storage reservoirs built in part to protect Mexican interests. By 1920, influential valley leaders such as Phil Swing, long an opponent of storage and a future congressman, recognized that the canal and storage were linked. Valley support for the 1922 Colorado River Compact, which divided the river equally between the upper and lower basins, and the construction of Hoover Dam by the Bureau of Reclamation, were the price that the valley had to pay for federal construction of an upstream storage facility and the All American Canal.

Eventually, all water rights and distribution systems were consolidated within the Imperial Irrigation District, formed in 1911 as the result of the bankruptcy of both the California Development Company and its Mexican subsidiary. Today, the district has rights to well over one-half of California's Colorado River entitlement, but its use is vulnerable to economic and political pressures for reallocation. The IID is the major remaining agricultural island in urban Southern California and a much coveted source of water for the MWD and the Los Angeles Department of Water and Power, who face unabated population growth and the loss of former supplies to Arizona and to other claimants in California. Los Angeles' Mono and Owens basin supplies have both been reduced, and MWD cannot count on the availability of surplus Colorado River water now that Arizona has begun to take its share through the Central Arizona Project.

The Imperial Irrigation District diverts its water directly from the Colorado at Imperial Dam above Yuma, where it enters the All American Canal. The water is used to generate power as it travels. Once the district makes a call on the Bureau of Reclamation for a delivery from Hoover Dam to Imperial Valley, it must move the water through its canals and laterals to the farmer. The district has five regulating reservoirs to conserve the water, but the total storage space is only 1,900 acre-feet (2,340 ML).

LEGAL BACKGROUND

Water rights on the Colorado River are defined by state and federal law. In 1928, Congress apportioned 7.5 million acre-feet (9.3 million ML) of the lower basin's share of the Colorado River among Arizona, California, and Nevada. California was given 4.4 million acre-feet (5.4 million ML); Arizona, 2.8 million (3.5 million ML); and Nevada, 300,000 (370,000 ML). In order to begin construction of Hoover Dam and the All American Canal, the seven major users in Southern

The patchwork of lush cropland near the Salton Sea in this 1965 photo from Gemini V shows how irrigation can "make the desert bloom." First called the Salton Sink, and later the Valley of the Dead, the region was renamed the Imperial Valley at the turn of the century as part of an effort to attract settlers to the harsh desert. When irrigated, the flat fertile land is tremendously productive, producing melons, lettuce, and other vegetables valued at nearly $460 million in 1988. CREDIT: L. G. Cooper and C. Conrad, Bureau of Reclamation.

California agreed among themselves on the allocation of the 4.4 million acre-feet (5.4 million ML). When the lower Colorado was adjudicated in *Arizona* v. *California* (1963), the Imperial Valley was given the right to divert the lesser of 2.6 million acre-feet (3.2 million ML) or an amount necessary to irrigate 424,125 acres as present perfected rights. Under the 1931 Seven Party Agreement, IID, along with the Coachella Water District and a small amount of Palo Verde acreage, has the third priority behind the Palo Verde Irrigation District and the California Division of the Yuma Project, but ahead of MWD, which has the fourth priority. Together, the first three priorities have a right to 3,850,000 acre-feet (4,749,000 ML) per year. The IID can also take an additional 300,000 acre-feet (370,050 ML) per year as a sixth priority if neither the Palo Verde Irrigation District nor the Yuma Irrigation District uses its entitlement.

The IID's water rights are subject to the requirement that the water be put to beneficial use. Western water rights are usufructuary—they are limited to the use of the water and cease if the water stops being put to a beneficial use. The prevailing assumption is that water that is not put to beneficial use is either forfeited or abandoned and is open to appropriation by others. Attacks on nonbeneficial or wasteful uses generally are brought by junior water rights holders against seniors. The IID has long been attacked for applying water in excess of crop needs, but its use was not seriously challenged until 1980. The most significant challenge to IID's water use was an ultimately unsuccessful effort by a valley activist to apply the excess land provisions of the Reclamation Act of 1902.

The geography of the valley makes excess use and the resulting drainage a serious problem. The U.S. Supreme Court Special Master, Simon Rifkind, noted in 1960 that much of California's water use was "wasted, as is apparent, for example, in the very large unused runoff each year into the Salton Sea." The IID has been sued by a number of landowners for damages caused by flooding. In 1980 a lawsuit filed by a Salton Sea farmer, who claimed that IID's tailwater drainage was raising the level of the Salton Sea and flooding his land, triggered a series of administrative and judicial orders requiring greater water conservation in the IID, although water salvage investigations had been under way since the mid-1960s. These state orders provided a strong incentive to negotiate a settlement (Reisner and Bates, 1990). In 1984, after a lengthy series of hearings in which a variety of interests (e.g., the California Department of Water Resources and the Environmental Defense Fund) presented substantial evidence on IID's water management practices and the potential for a conservation-induced water sale, the State Water Resources Control Board concluded that IID's use of water was unreasonable under California law and constituted waste (Stavins and Willey, 1983). The next year the Bureau of Reclamation issued a report identifying measures that could conserve 354,000 acre-feet (436,700 ML) per year.

THE IID-TO-MWD TRANSFER NEGOTIATIONS

The 1989 Water Conservation Agreement signed by IID and MWD accomplishes the largest recorded transfer of salvaged water between an agricultural user and an urban supplier. It is important both in and of itself and as a harbinger of future transfers. The transfer came as a result of both market incentives and legal sanctions. All the major interests—IID, other irrigation districts with higher priorities, MWD, and national environmental organizations, some of whom actively encouraged the transfer—describe it as a "win-win" agreement. The possible third party effects directly attributable to the 100,000-acre-foot (123,400-ML) transfer seem to have been addressed adequately.

Four rounds of negotiations were necessary to arrive at the current agreement. The IID initially proposed to sell water to MWD at about $250 per acre-foot ($203 per ML) with an inflation adjustment; IID pays the Bureau of Reclamation between $10 and $30 per acre-foot ($8 and $24 per ML). At the same time, the district was challenging the state board's authority to impose conservation measures. The sale proposal was controversial for both legal and political reasons. The legal problem was that all participants in the 1931 Seven Party Agreement, including IID, took the position that water not used by a senior passed to the next lower priority user who could put it to beneficial use. Simply put, any wasted water was not IID's to sell but MWD's to take.

The Imperial Irrigation District broke with this principle, and MWD dropped its junior entitlement claims. The parties agreed that MWD would pay $10 million into IID's conservation fund in return for just over 100,000 acre-feet (123,400 ML) for 35 years. Negotiations were broken off after IID's board rejected the agreement by a 3 to 2 vote in late 1985 and MWD refused to consider a transfer that allowed the district to receive a large profit over and above the capital and maintenance costs of the conservation regime. For a time, an IID consultant urged the district to privatize the transaction. Direct negotiations between IID and MWD resumed, however, in 1987 after an appellate court held that the state had the authority to order the district to institute a water conservation program (*Imperial Irrigation District* v. *State Water Resources Control Board*, 1986).

The two parties reached a tentative agreement in 1987, but IID continued to adhere to its position that it was selling water. In early 1988, IID dropped the price to $175 per acre-foot ($142 per ML), but MWD refused to counter. As the State Water Resources Control Board proceeded to hold hearings on the means of compliance with a target of 350,000 acre-feet (432,000 ML) of conserved water per year, IID became more flexible. Between May and November of 1989, IID and MWD worked out details of the final agreement, after the parties agreed to split the difference on their estimates of the value of the water, bringing the figure to around $100 per acre-foot ($81 per ML).

THE 1989 WATER CONSERVATION AGREEMENTS

The 1989 Water Conservation Agreement obligates MWD to pay for both structural and nonstructural conservation projects designed to conserve 106,100 acre-feet (130,900 ML) annually, which it can then take for a period of 35 years. Total capital costs are estimated to be $97.8 million, plus $23 million in indirect costs. The program will be implemented over a 5-year period between 1990 and 1994; then the savings will be constant until MWD or another party agrees to an additional conservation program. The legal questions that initially created potential barriers were finessed. Section 6.5 of the agreement is replete with disclaimers that MWD shall not assert a right to the conserved waters, that the rights of the parties "except as specifically set forth in this agreement" to the use of the Colorado are not affected, and that the conserved water will at all times retain its third priority and has not been forfeited by IID. Likewise, the possibility of MWD's banking water in Lake Mead during wet years is acknowledged, but the irrigation districts reserve the right to challenge any future banking agreements.

Canal lining is the major conservation strategy being pursued, but IID will also build new regulating reservoirs and canal spill interceptors and will automate its delivery system. The agreement will be administered and monitored by a program coordinating committee composed of one IID, one MWD, and one neutral representative. The MWD must bear all capital construction costs, the "ongoing direct costs of the nonstructural projects of the program, and operation, maintenance and replacement costs of the structural projects of the program necessary to the keep the projects . . . in good operating condition during the terms of this agreement" (IID and MWD, 1989). In addition, MWD must bear $23 million in indirect costs such as lost hydroelectric revenues, "mitigation of adverse impacts on agriculture from increased salinity in the water," and environmental mitigation and litigation costs from any impact on water levels or water quality in the Salton Sea and the New and Alamo rivers (IID and MWD, 1989).

In return for financing the conservation program, IID agrees to reduce its requests to the Secretary of the Interior for water from the Colorado River "below that which it would otherwise have been absent the projects of the Program (in an amount equal to the quantity of water conserved by the Program. . . ." (In normal circumstances, IID will reduce its diversions by 106,100 acre-feet (130,900 ML) annually.) The MWD can make a corresponding additional request of the Secretary of the Interior at its Lake Havasu intake for the amount saved. The agreement also addresses how the burden of shortages will be shared.

The IID and MWD Water Conservation Agreement is seen by both parties as a win-win agreement. The MWD, of course, increases its Colorado River supplies by 106,100 acre-feet (130,900 ML), and this additional margin of safety, about 20 percent of its anticipated dry year shortfall, will become important as Arizona takes more of its Colorado River entitlement and supplies become more scarce in both the Colorado basin and northern California. The IID obtains money in return for doing what it may well have become legally obligated to do in any event, while its irrigated acreage is unaffected by the agreement. This is the first such transaction and thus relatively straightforward to negotiate. Future attempts to reach such agreements may be more complicated.

THIRD PARTY IMPACTS

The three major third party claimants are two other irrigation districts, Palo Verde and Coachella Valley, and the public environmental values associated with the Salton Sea. Future lining of the All American Canal would affect Mexican irrigators who depend on seepage from the All American Canal but were not acknowledged as having a legal entitlement to such water in the treaty between the United States and Mexico allocating the water of the Colorado between the two nations.

The Coachella Valley Water District lies northwest of the Imperial Valley. It is a relatively more efficient water user and is concerned that the 1989 Water Conservation Agreement will decrease the return flows it receives from IID. Palo Verde is upstream from IID's intake but below MWD's, so any water taken by MWD can affect Palo Verde. After the 1989 agreement, Coachella filed suit against IID and MWD, basically arguing that under the Seven Party Agreement conserved water must first be shared with other agricultural users. In December 1989 the Coachella Valley Water District, Palo Verde Irrigation District, and IID settled the lawsuit by an approval agreement. To ensure that MWD is taking water on a one-for-one basis for conserved water, the agreement created a measurement committee to verify that IID's programs do in fact conserve water. To resolve the concerns of the Coachella Valley Water District, MWD and IID also agreed to provide protection for Coachella during extremely dry periods when Colorado River supplies fall to critical levels. This protection, if implemented, would reduce the availability of water to MWD during these rare hydrologic events. To compensate for this,

MWD will receive 106,100 acre-feet (130,900 ML) annually instead of the original 100,000 acre-feet (123,350 ML) agreed to in 1985. In addition, the minimum agreement period (currently 35 years) will be extended by 2 years for each year that MWD's supplies are reduced.

The major environmental values most directly affected by the 1989 agreement are those associated with the Salton Sea. The Salton Sea is 278 ft (93 m) below sea level and was formed when a new channel was opened by the great Colorado River flood of 1905. The conservation measures will lower the level of the lake by about 2 ft (0.67 m) over the life of the agreement. One environmental impact is that lower lake levels, which benefit littoral lands, also result in increased salinity levels in the lake.

The Salton Sea also is threatened by the New River, which is basically a conduit for untreated sewage, pesticide residues, and heavy-metal contamination from northern Mexico. Pollution is a serious problem along large stretches of the United States-Mexican border. The problems have not been addressed adequately either in Mexico or through international organizations, although the pending Mexico-United States Free Trade Agreement may provide some impetus for the two nations to improve environmental conditions along the border (U.S. EPA, 1991). Such problems are beyond the scope of the IID-MWD agreement, but their existence cautions that even though an individual reallocation may be judged positive, that judgment may be qualified when the case is viewed in a broader context.

FUTURE AGREEMENTS

The Metropolitan Water District of Southern California has engaged in other transactions to conserve, store, and transfer water. Moreover, such arrangements are being pursued by MWD for the future. The following is a brief description of other similar activities being undertaken by MWD.

1. Arvin-Edison/MWD Exchange. In the San Joaquin Valley, MWD has developed a program with the Arvin-Edison Water Storage District, a large federal Central Valley Project (CVP) contractor, for the storage and transfer of water. Under this program, MWD receives dry year CVP supplies that would otherwise be used by Arvin-Edison in exchange for State Water Project (SWP) supplies previously delivered to Arvin-Edison during wet periods. When MWD withdraws water from its storage account, Arvin-Edison would pump up and deliver the previously stored SWP water to the farmers in its service area.

The program is expected to increase reliable supplies available to MWD by about 93,000 acre-feet (115,000 ML) annually under conditions similar to the 1928 to 1934 drought, while improving the local agricultural economy. Implementation of the program, which is now in the final stages of the environmental documentation process, will require capital expenditures within Arvin-Edison of about $20 million for expanded spreading works, a distribution system, and increased ground water extraction capacity. The unit cost of the program is about $90 per acre-foot ($73 per ML).

2. Desert-Coachella Exchange. Under an agreement initially negotiated in 1967, MWD provides additional Colorado River water to ground water basins serving Coachella and the Desert Water Agency. In exchange, MWD can receive during dry periods more than 60,000 acre-feet (74,000 ML) of SWP entitlement water paid for by these other agencies. By April 1986, MWD had accumulated a storage account of 552,000 acre-feet (680,900 ML). During the ongoing drought, MWD has stopped delivery of water to the exchange and drawn down its storage account to about 420,000 acre-feet (518,000 ML).

3. Palo Verde Water Utilization Agreement. Beginning in 1986, MWD conducted negotiations with Palo Verde Valley landowners and the Palo Verde Irrigation District, which has the most senior rights to Colorado River water. The purpose of the negotiations was to reduce the amount of irrigated land in the Palo Verde Valley to make an additional 100,000 acre-feet (123,400 ML) of water available to MWD in dry years. Discussions with the Palo Verde Irrigation District have resumed, following the finalization of negotiations on the Imperial Conservation Program.

4. Department of Water Resources Activities. In addition to short-term water purchases from the Yuba County Water Agency and from La Hacienda, Inc., in Kern County, the Department of Water Resources continues to explore water transfers as a means of increasing the long-term yield of the State Water Project. These activities include: (a) negotiations with the Yuba County Water Agency for a long-term water transfer supply, (b) development of the Kern Water Bank, and (c) possible conjunctive use programs with other Central Valley agencies to increase available supplies to SWP contractors. Part of the yield of this innovative and complex conjunctive use program in Kern County will require a transfer of State Water Project entitlement water from the Kern County Water Agency to the other SWP contractors in exchange for use of the water previously stored underground in Kern County by the Department of Water Resources.

5. Future MWD Transfer Activities. The MWD intends to continue to identify and develop water transfer programs in the future. These programs will emphasize the use of financial incentives in agricultural areas to increase conservation and improve water management, thereby making additional water available to meet southern California's needs. Future programs will include (a) IID-type conservation programs, especially where technically and politically feasible in the drainage-impacted portions of the western San Joaquin Valley; (b) conjunctive use programs similar to the Arvin-Edison exchange; and (c) agreements with landowners and their water agencies to alter farming practices, for example, by fallowing some land in their crop rotation or implementing on-farm conservation, to make additional water available for use by growing urban areas (Quinn, 1990).

6. Other MWD-IID Agreements. The Imperial Irrigation District and the Metropolitan Water District are also currently in negotiation regarding another agreement that would operate in much the same way as the initial 1989 Water Conservation Agreement. The MWD would pay for the implementation of conservation measures. It would then be entitled to place an additional request for water from the Colorado River corresponding to the amount salvaged. Implementation of these conservation measures would result in an additional 150,000 acre-feet (185,000 ML) of conserved water. It is anticipated that the Coachella Valley and Palo Verde districts will insist on measures to protect their interests. Furthermore, closer scrutiny is anticipated from environmental interests concerned about adverse environmental effects, particularly with regard to the increased salinity of the Salton Sea. Many environmental interests support water transfers as a means of avoiding the environmental costs associated with water projects. Indeed, the Environmental Defense Fund played an important role in negotiations leading to the 1989 Water Conservation Agreement between IID and MWD. However, heightened concerns about exacerbating the problems of increased salinity levels in the Salton Sea might limit such support.

In 1988, Congress passed legislation authorizing the lining of portions of the All American Canal and the Coachella branch of the All American Canal. All costs would be paid by the California agencies receiving the salvaged water. The lining project is expected to save 100,000 acre-feet (123,350 ML) annually. An experimental project is being conducted by the Bureau of Reclamation to develop new techniques to line canals while water continues to flow through them.

The Imperial Irrigation District has indicated an interest in paying for the lining of the canal, apparently in anticipation that it would then be in a position to market the salvaged water. However, MWD takes the position that such a plan would violate the priorities identified in the Seven Party Agreement, which allocates Colorado River water among the California agencies that rely on it. A further complication exists because Mexican farmers rely on water that seeps from the All American Canal. Although they apparently have no legal right to such water, potential adverse third party impacts on these Mexican communities could be significant. Further, environmental values such as the wetlands created by the leaks may also be harmed.

CONCLUSIONS

The Imperial Valley case study illustrates four conditions that are likely to produce a presumptively win-win transfer: (1) a large user with surplus water as a result of past practices, (2) the threat of judicial or administrative reallocation due to alleged waste, (3) a well-financed willing buyer with few reasonably priced alternative supplies, and (4) the paucity of well-defined external costs from the specific transfer. The Imperial Valley is beset with several serious environmental problems such as border pollution and the fate of the Salton Sea. The transfer worked, in part, because it was relatively neutral with respect to these problems. When the four conditions are not present in this combination, it will be harder to negotiate win-win transfers. Still, the Imperial Valley shows that a market transaction, stimulated by the threat of judicial or administrative reallocation, can move water to an area of higher demand.

REFERENCES

Arizona v. California, 373 U.S. 546 (1963).

Hundley, N. 1975. Water and the West: The Colorado River Compact and the Politics of Water in the American West. Berkeley: University of California Press.

Imperial Irrigation District v. State Water Resources Control Board, 186 Cal. App. 3d 1160 (1986).

Imperial Irrigation District v. State Water Resources Control Board, 275 Cal. Rptr. 250, 267 California Court of Appeals, 4th District (1991).

Imperial Irrigation District (IID) and Metropolitan Water District of Southern California (MWD). 1989. Agreement for the Implementation of a Water Conservation Program and Use of Conserved Water. December.

Quinn, T. 1990. Shifting water to urban uses: Activities of the Metropolitan Water District of Southern California. Paper presented at the Natural Resources Law Center, Boulder, Colo., June.

Reisner, M. 1986. Pp. 125-136 in Cadillac Desert: The American West and Its Disappearing Water. New York: Viking.

Reisner, M., and S. Bates. 1990. Overtapped Oasis: Reform or Revolution in Western Water Law? Washington, D.C.: Island Press. 196 pp.

Stavins, R., and Z. Willey. 1983. Trading Conservation Investments for Water: A Proposal for the Metropolitan Water District of Southern California to Obtain Additional Colorado River Water by Financing Water Conservation Investments for the Imperial Irrigation District. Environmental Defense Fund, Berkeley, Calif.

U.S. Environmental Protection Agency (U.S. EPA). 1991. Integrated Environmental Plan for the Mexico-U.S. Border Area (first stage, 1992-1994). USEPA Working Draft, August 1.

Worster, D. 1985. Pp. 194-212 in Rivers of Empire: Water, Aridity, and the Growth of the American West. New York: Pantheon Books.

12

Conclusions and Recommendations

The West's water needs are changing. Rapidly increasing economic and population growth in urban areas has generated corresponding increases in demand for augmentation of water supplies. Irrigation, by far the largest water use, remains a mainstay of some local and state economies. Perhaps the most rapidly escalating call for water is motivated by concern for environmental and recreational values, values not protected by law or public advocacy in the early evolution of western water allocation. These increasing and shifting patterns of demand are being exerted on a resource already fully appropriated in most of the region.

The committee believes that voluntary water transfers are the single most significant tool available for responding to these new and changing water needs. It is nevertheless the case that transfers sometimes are proposed without proper regard for third party interests. Based on its review of the potentially adverse impacts of water transfers, this committee concludes that third party interests deserve greater consideration when transfers are proposed. While seeking ways to promote transfers, state and tribal governments should also devise ways to improve their laws and procedures to protect third parties. Federal agencies involved in water allocation and management should also promote transfers and protect third party interests in the process. Each sovereign must devise its own specific approach, suited to its own objectives, but the committee offers recommendations to aid decisionmakers in designing these approaches. The recommendations are based on the fundamental premise that transfers should meet the needs of a changing West while causing a minimum of adverse impacts.

The committee examined actual and potential water transfers in diverse areas of the American West to understand better the nature, scope, impacts, and institutional setting of water transfers. Every situation is unique, but there are important commonalities. The committee believes that the problems and opportunities illustrated in the various case studies are not unique and will arise elsewhere throughout the West. Thus valuable lessons for future federal and state water allocation policy can be drawn from the studies.

CRITICAL ISSUES

Three critical issues received recurring attention during the course of this study—area-of-origin protection, instream uses, and transaction costs—and should be considered key areas of concern for federal, state, and tribal governments committed to improving their water transfer processes.

Area-of-Origin Protection

If water transfers are to be used to facilitate more responsive water management, the equity issues related to area-of-origin impacts will require continued attention. Although an individual farmer might benefit from selling water rights to satisfy growing urban demand, rural counties are left with the problem of trying to protect their tax bases, environments, cultures, and economic futures. The Arizona Ground Water Management Act, as enacted in 1980, is one example of this dilemma. It allowed water to be transferred from agricultural areas to municipal and industrial use, but it has no mechanisms for considering the interests and values of the areas where the water originates. Changes enacted in 1991 do provide some area-of-origin protection.

Similar concerns are expressed by people in communities where social and cultural values may depend on protecting existing water uses and keeping water in natural water courses and wetlands. They fear that out-of-basin transfers will undermine the integrity of the communities themselves. In northern New Mexico, for example, the seemingly neutral institutions of adjudication—which can include a public interest review of new appropriations and transfers—do not fully reflect the concerns of the Hispanic community. Lifestyle, community organization, and personal relationships are all intimately related to traditional water uses and allocation arrangements. The

New Mexico state system, however, like most western states, regards water essentially as a property right and a commodity; it does not see water as an element of community cohesion, history, and collective aspirations.

Instream Uses

It is also clear that one of the major issues for decisionmakers is the balancing of consumptive and nonconsumptive, or instream, uses. Prior appropriation allows private rights to be created in public resources, and throughout the West's history, appropriative water rights were typically held by private parties or public water providers for consumptive uses. (An exception was made for hydropower rights, which could be exercised only by building facilities in the stream to take advantage of flowing water.) However, many of today's demands are for nonconsumptive uses that require flowing water to be left in the stream. With few exceptions (e.g., New Mexico), virtually all western states have enacted programs that legally recognize instream uses. However, because water in most western streams is fully appropriated, such mechanisms face significant constraints. In most cases, a public entity holds such rights and they are relatively junior in priority.

Several of the case studies illustrate that transfers can be used to help satisfy demands for instream flows with more reliable senior rights. However, existing institutions generally do not encourage transfers for nonconsumptive uses and may in fact inhibit them. The Truckee-Carson experience illustrates the positive potential of agriculture-to-wildlife and wetland transfers. Incentives for transfers in this case included both the threat of a judicially mandated involuntary transfer and the provision of funds to induce water rights sales. Voluntary transfers to instream uses have not occurred in other areas, such as the Central Valley and the Yakima Valley, where courts have not intervened or where they have confirmed existing uses.

Transaction Costs

The potential for increased transaction costs caused by greater consideration of third party effects must be addressed. Public policy has been greatly influenced by the concept of transaction costs and the need to minimize them. Transaction costs are the costs of negotiating and enforcing transfer agreements and clarifying property rights so that transactions may proceed. The higher the transaction costs, the lower are the profits or economic benefits that accrue to those transferring the water from one use to another, and therefore the lower are the incentives motivating transfers.

Some transaction costs are a necessary part of enacting any water transfer. Others, however, result from ambiguous policies or criteria for transfer approval, and from costly and duplicative efforts to provide information about a transfer and its potential impacts. Additional costs are incurred when the range of parties engaged in the transfer process is expanded to include third parties.

In general, transaction costs that are incurred in empowering and accommodating traditionally underrepresented third party interests are beneficial. In other words, society agrees to incur costs beyond those basic to the process of buying and selling water in the hope that its values will be more broadly represented and the hidden costs of transfers addressed. As interest in water transfers as a voluntary reallocation tool has increased, there has been a growing call to streamline the process to encourage market transfers. The case studies reviewed in this report suggest, however, that this goal must be tempered by the reality that some increased transaction costs are necessary if we are to address third party effects adequately. Although substantial opportunities exist to improve the transfer procedures in use in the West and reduce some costs, transferring water is no simple matter. Thus no simple and inexpensive process will be able to meet the needs of buyers, sellers, governing bodies, and affected third parties equitably.

CONCLUSIONS AND RECOMMENDATIONS

Conclusion 1: Water transfers can promote the efficient reallocation of water while protecting other water-dependent values recognized by society.

Changes in the use of water—whether in the point of diversion, location of use, or type of use, and whether or not accompanied by changes in ownership of water rights—are necessary and desirable in a dynamic society. Changes in use often are driven by changes in the perceived economic value of a particular water use. For most of the West's history since settlement, diversions have been from natural watercourses and instream uses to storage and off-stream use, initially for mining, later for agricultural and municipal purposes. These demands led to the construction of major storage and conveyance facilities, including systems for interbasin transfers within and among states.

Economic reality and environmental concerns limit the construction of additional water development projects designed to increase supplies. As demand approaches the limits of available supplies, the movement of currently available water from relatively low-valued to relatively high-valued uses becomes an increasingly attractive alternative. Furthermore, the water transfer agenda now includes acquisition of water rights and changes in water project operations to restore and protect instream flows and associated values that were degraded by past patterns of water development and use.

The committee supports water transfers as one component of efficient water management, provided that such transfers are accomplished equitably. One problem is that existing laws, policies, and procedures concerning water market transactions and other transfers often fail to ensure either that third parties are protected from negative effects or that they share the benefits. Affected parties can include existing rights holders, rural communities, unique cultures, the environment, and other interests beyond the willing seller and willing buyer. The impacts can be obvious—increased per capita costs for irrigation system maintenance and operation, and loss of county tax revenues—or subtle—the diminished viability of rural economies, a loss of confidence in the community's future, and the erosion of unique cultural values of water-dependent communities. Examples of possible environmental impacts include instream flow losses and water quality alterations affecting fish and wildlife; changes in aquatic and riparian habitats, stream channel integrity, and esthetic values; and the loss of recreational uses.

The "no injury" rule that historically has governed water transfers under the prior appropriation doctrine in most western states is the foundation for third party protection, but it generally is not adequate to protect the full array of affected interests. Although many states have adopted a public interest review requirement for proposed transfers, others still rely exclusively on a no injury test that protects only other water rights holders. Further, public interest requirements in some states are not clearly defined, leaving state administrators with little guidance on how to apply the policy. Many changes in state and federal laws, policies, and procedures are needed to provide appropriate protection for the full range of water uses and users, natural environments, basins of origin, unique cultures, and communities.

RECOMMENDATION:

- All levels of government should recognize the potential usefulness of water transfers as a means of responding to changing demands for use of water resources and should facilitate voluntary water transfers as a component of policies for overall water allocation and management, subject to processes designed to protect well-defined third party interests.

Conclusion 2: State and tribal governments have primary authority and responsibility for enabling and regulating water transfers, including identification and appropriate mitigation of third party effects.

States historically have had primary authority in the administration of water rights, except on Indian lands where tribal governments are the administrative authority. Thus the administration of water transfers is the responsibility of the state or tribal governments. These governments are capable of assessing transfers in the context of the region's water management needs and of providing accessible mechanisms for revealing and addressing third party impacts. This authority is extensive, including transfers of federal project water. Federal intervention may be necessary, however, when there is an overriding national interest such as interstate relations, navigation, or endangered species.

State and tribal governments should design administrative processes so that the scale of regulation reflects the scale of effects. For example, transfers of water rights within a basin involving no change in use and no change in point of diversion should be processed with minimal delays and procedures while still meeting baseline public notice, protest, and hearing requirements. The most straightforward transfers—those involving minor and noncontroversial changes of use within a basin—could be treated as presumptively approvable as long as direct third party effects are identified and reasonable mitigation measures are presented in the transfer application. The burden of proof for disapproval would fall on regulators and opponents.

Large or complex transfers within basins, most interbasin transfers, and interstate transfers need additional scrutiny. These should be evaluated against public interest criteria defined by the state, with consideration of biological, physical, and economic efficiencies, and of significant economic, environmental, and social effects on third parties. The burden of ensuring that third party impacts are revealed and considered in the review process should be shared by the transferring parties and state and tribal governments.

Recognition of rights to instream flows is one way of preserving environmental interests within state and tribal water law systems.

State laws regarding the acquisition of water rights for instream flow purposes vary. Some states allow any party to acquire and hold rights for instream flow purposes, whereas others require such rights to be held by a designated state agency. Given the importance of instream flows to a variety of third parties, authority to acquire and hold water rights for instream flow purposes should be a feature of water law and administrative practice in all states.

RECOMMENDATIONS:

- State and tribal administrators should develop and publish clear criteria and guidelines for evaluating water transfer proposals and addressing potential third party effects.
- State and tribal administrative processes should provide for public and broad third party representation in the review of water transfer proposals. In addition to normal actions such as notices of proceedings, public hearings, and protest opportunities, programs should include affirmative review of potential third party effects in cases likely to involve significant effects.
- State and tribal processes should seek to regulate water rights transfers in ways appropriate to the scale of effects with the dual objectives of avoiding excessive transaction costs and providing meaningful consideration of third party interests.
- State laws should allow governmental entities to acquire water rights for instream flow purposes with the same priority and protection against injury enjoyed by rights held for other uses. Affirmative state policies may be necessary and appropriate to acquire sufficient instream flow rights to mitigate the effects of historic diminutions of streamflow.
- States should provide leadership in exercising their water administration and planning responsibilities to identify opportunities for water transfers that might serve as instruments for achieving a wide range of water management objectives.

Conclusion 3: Water transfer law and policies should be designed to consider the interests of the trading partners, third parties, and the environment in a cost-effective manner.

In designing transfer policies, decisionmakers face serious tradeoffs between the desire to preserve the direct private or public gains from water transfers and the desire to protect third party interests. Such policies significantly affect the relative benefits and costs of proposed transfers and the distribution of these benefits and costs. As a general rule, regulatory processes and requirements that attempt to protect against all third party impacts regardless of the nature or magnitude of the effect or the standing of the party result in high transactions cost and discourage desirable transfers. Conversely, procedures and processes that ignore third party interests altogether allow transfers to impose inequitably large losses on third parties. Either result is inefficient and inequitable. Regulatory processes and requirements can distinguish between large or pervasive third party effects and small or ephemeral impacts. In many instances the costs of accommodating or mitigating third party impacts can be defrayed from the economic benefits that the transfer generates. Transaction costs can be reduced by taking steps to improve the information base that is used to evaluate a transfer, encouraging negotiated resolutions to conflicts, and establishing streamlined procedures for estimating transferable quantities based on consumptive use for various categories of use.

Many disputes between transfer applicants and objectors involve the quantity of water that should be transferable. Policies that presume a transferable quantity per irrigated acre may help reduce transfer costs. New Mexico, for instance, sets a standard quantity of water that may be transferred per unit of irrigated land retired. Parties who disagree with this quantity bear the cost of proving that some other amount is appropriate. This system, however, requires third parties to take on a potentially difficult and expensive task—quantifying the amount of water that can be transferred. In contrast, the Colorado transfer review process requires the transfer applicant and all objectors to provide evidence on transferable quantity, thus possibly incurring higher cumulative costs for engineering and legal studies but providing more protection. Each state can consider ways to reduce the informational costs of assessing transfer impacts as well as the costs created by ambiguous criteria.

Litigation often is inevitable to resolve water use conflicts, but adversarial proceedings are not a good forum to explore a wide range of options affecting multiple interests. Nonjudicial conflict resolution processes might be valuable for providing comprehensive evaluation of transfer proposals. There is a need to allow others besides attorneys and technical experts to share in framing the issues, as well as a need to share the information among all interested parties.

Transaction costs incurred to address third party concerns serve a beneficial role when they give affected third parties a voice in the review process. However, there are many opportunities to reduce unnecessary transaction costs by clarifying state policies and by encouraging actions to lower the costs of gathering information on potential impacts.

RECOMMENDATIONS:

- The costs of mitigating third party effects should be internalized as a cost of the transfer—that is, the beneficiaries or proponents of the transfer should bear the mitigation costs as a matter of law and equity. Therefore the cost of the transfer should include sufficient funds to help mitigate third party effects, in the form of water, money, or other compensation.
- To help reduce costs, policies might be designed so that, in general, transfers of acquired rights are limited to consumptive uses. This may entail setting state, river basin, or regional standards for the consumptive use of water per irrigated acre based on crop type, historic water availability, and other local variables. Such standards should be flexible enough to account for variations in water availability and local conditions. Third parties should not have to develop data on the transferable quantity; data should be developed by the buyer or seller.
- Regulatory requirements should be designed to encourage negotiated resolutions of conflicts. Consideration should be given to processes other than judicial proceedings (e.g., a state water court) to provide the initial evaluation of transfer proposals.

Conclusion 4: Water transfers between basins should be evaluated to determine and account for the special impacts on interests in the areas of origin.

Water is not merely a commodity in the normal sense of the word but rather a resource held in common for all citizens, and this should be recognized in the processes used to evaluate water transfers. The interests of communities, local governments, individuals, and the environment in the areas from which water is proposed to be exported are not now adequately represented in the laws and policies of many western states. Communities hold strong values that support maintaining water within its natural watershed. For some, water is the basis of present and future economic vitality and environmental amenities; these values often are augmented by water-related social and cultural concerns and traditions. Among the cases studied, the threat to cultural values is illustrated dramatically in the case of the acequias of northern New Mexico.

State laws and policies often fail to recognize that when water is removed from its natural watershed a variety of economic, social, and environmental harms can occur. Prior appropriation law has never limited the use of water to the watershed in which it originates, and the committee does not recommend that water be constrained to its watershed of origin. However, states do need to ensure that the special problems caused the public by transbasin export are fully addressed before such transfers are permitted. Although basin-of-origin interests should not have veto power over transfers, their interests should be represented in the evaluation process, to the extent that those interests are important to society.

State tax laws often provide incentives for transfers because local taxing jurisdictions cannot impose taxes or in lieu payments, and these incentives could be revised to address the equities of reallocation. Municipalities and certain quasi-public water utilities that purchase or lease water supplies or build and operate works to use water supplies from outside their service areas ordinarily pay no property taxes to local governments in areas of origin because of intergovernmental immunity from taxation. In addition, western states usually provide by law that private entities may be taxed on the value of water facilities only by the county where the water is used. The inability to tax facilities that dewater areas of origin compounds the disadvantages to the communities in these areas. For instance, states could revise their tax laws to make exporting entities bear a larger portion of the costs of mitigating the third party effects of transfers. Exporters could be subject to local property or other taxes. States might consider developing a transaction tax for water rights transfers. This transaction tax would be analogous to a severance tax and would be used to mitigate adverse environmental effects caused by transfers (interbasin as well as others). Setting the level of such a tax would require careful analysis so that desirable transfers are not discouraged but adequate revenues for mitigation are collected.

RECOMMENDATIONS:

- States and tribal governments should develop specific policies to guide water transfer approval processes regarding the community and environmental consequences of transferring water from one basin to another, because such transfers may have serious long-term consequences.
- Water transfer processes should formally recognize interests within basins of origin that are of statewide and regional importance, and these interests should be weighed when transbasin exports are being considered.

• Although each state or tribe should select the approach that suits its needs best, area-of-origin protection generally would include impact assessment, opportunities for all affected interests to be heard, regulatory mechanisms to help avoid adverse effects, compensation (e.g., financial payments or mitigation), and authority to deny a proposed transfer or water use involving a transbasin export if the effects are judged unacceptable.

• States should revise laws that now exempt water facilities from taxation by the county of origin either because the exporter is a public entity or because of provisions that make such facilities taxable only in the county where the water is used. Mechanisms to compensate communities for transfer-related losses of tax base, such as an annual payment in lieu of taxes, may be needed.

Conclusion 5: Public interest considerations should be included among the third party issues and legal provisions for permitting, conditioning, and denying water transfers.

The committee recognizes the difficulties inherent in representing and quantifying public interest values in state and tribal forums. Many state laws, for example, direct administrators to consider the public interest as they review proposed transfers, but few provide real guidance as to what constitutes the public interest or offer specific procedures to ensure probable effects and adequate representation of the public interest. Legal inadequacies and uncertainties often restrict or hinder economically efficient and environmentally desirable water rights transfers. These uncertainties and problems also often interfere with effective actions on behalf of third party interests, including factors usually referred to as "public interest" and "public trust."

Although this report makes a modest attempt to suggest some elements critical to any definition of the public interest, state and tribal governments ultimately must provide such guidance for their water administrators. This guidance must include criteria to be considered in a public interest review and also procedures to provide for adequate representation of the interests.

RECOMMENDATIONS:

• To protect third parties in water rights transfers, public interest language in western states' water laws should be reviewed, clarified, and, where appropriate, more vigorously applied. States should develop definitions and criteria for assessing what constitutes the public interest, perhaps benefiting from the legislative and judicial initiatives developed by the states of Idaho and Alaska. Such definitions should embrace existing water rights holders, environmental water needs for ecosystem protection, and social and cultural values in basins of origin.

• To the extent that public trust concepts and values cannot be represented dependably under existing laws and policies, states should develop new laws, institutions, and administrative tools for doing so. Key elements of public trust administration might include comprehensive planning at the river basin level, including the identification of existing social and environmental values dependent on water, clearly defined procedures to guide applicants seeking water transfer approval, and institutional arrangements for holding and managing water for instream and environmental uses.

Conclusion 6: Environmental impacts can and should be considered by state, tribal, and federal agencies when potential water transfers are evaluated.

Transfers of water may have significant effects on natural ecosystems, ranging from loss of aquatic and riparian habitats associated with dewatering streams to water quality effects arising because of increasing concentrations of pollutants. Direct impacts on plants and animals also are caused by the construction and operation of physical conveyance systems used in support of water transfers. Those states (such as Colorado) that rely solely or primarily on the prior appropriation system and a test of no injury to other water rights holders do not adequately assess injury to natural ecosystems. Environmental protection should be part of the basis for impact assessment and mitigation relating to proposed transfers.

Current law requires significant environmental studies when new structural water projects are undertaken, including reservoirs, channelization, power generation, and diversion projects. However, little effort is made to quantify or characterize the local and regional environmental impacts caused by water transfers and related conveyance systems. This inequity will need attention as water transfers become an increasing component of water management.

RECOMMENDATIONS:

• Federal, state, and tribal water transfer policies and laws should ensure consideration of ecological values affected by transfers, for example, the goals of protecting riparian and wetland habitats and the water needs of endangered species. Adjustment of water laws, administrative practices, and water supply strategies may be necessary to achieve this end.

- States should develop inventories of wetland and riparian systems that depend on surface water systems likely to be sought for transfers so that transfers can then be evaluated in light of ecosystem protection and restoration goals.
- When water is transferred to new uses in new areas, there is an opportunity to promote overall efficiency by dedicating some of the newly available supply to public uses such as environmental protection. Congress should consider using a share of the receipts gained from marketing federal project water to acquire water to protect western biota and habitats. The federal Land and Water Conservation Fund Act might be amended to permit both state and local grant recipients and federal agency participants to use fund money to acquire water rights for environmental protection and mitigation.

Conclusion 7: Traditional Indian and Hispanic communities have unique interests relating to water transfer policies, and these interests merit special consideration when proposed transfers are evaluated.

In many water transfers, traditional Indian and Hispanic communities find themselves in a classic third party role. Such communities view water rights administrative processes—including rights filings, adjudications, and transfers—as a threat to their historical uses of water. Also, they often view the prior appropriation doctrine as serving individual interests at the expense of community interests and as hostile toward communal social values.

Many Indian and Hispanic communities had well-established water use and allocation systems that preceded state systems. When states adopt the prior appropriation system and apply it to these communities, they can cause considerable disruption of longstanding water management processes. For example, in the acequias of northern New Mexico, prior appropriation adjudications imposed a system of individual priorities for the first time on a tradition that called for communal sharing of surpluses and shortages. The new priority system fractured longstanding and important communal structures.

It is important to note that traditional Indian and Hispanic communities were not adequately represented in state water adjudications because members lacked not only knowledge of the importance of the adjudications but also the resources to assert their interests fully. As a result, in many instances, individuals within the community were decreed no water or less water than had been used historically. Moreover, during the era of federal investment in Bureau of Reclamation water infrastructure and delivery systems, traditional communities typically were ignored. As a consequence, water supplies for these communities were limited. Thus, in evaluating proposed water transfers, states should consider options to supplement existing decrees that do not adequately address the historical water uses of Indian and Hispanic communities.

RECOMMENDATIONS:

- States should carefully scrutinize any proposed water transfer that could adversely affect water supplies for Indian and Hispanic communities.
- States should consider enacting legislation that permits the establishment of historical or cultural zones as a means of insulating Indian and Hispanic communities from further injury. Transfers that would have adverse impacts on these zones would receive strict scrutiny.

Conclusion 8: Tribal governments should consider special factors in approving and administering water transfers on their reservations.

Indian tribal governments play a dual role when water transfers involve reservation water. As proprietors, tribes typically have senior water rights within a state appropriation system; as sovereigns, tribes have authority to administer members', and often nonmembers', water use on the reservation. Senior reserved Indian water rights were essentially ignored during the era of federal investment in large irrigation storage and water delivery systems. As a consequence, Indian communities generally lack the facilities to use much of their water entitlement and often lack the capital to develop such facilities. In the meantime, non-Indian appropriators, often with the assistance of federal reclamation projects, have used water from the streams where Indian rights exist.

Tribal governments have on-reservation authority for administering water transfers. Setting tribal water transfer policies and administering them present special challenges. Tribes desire some flexibility to adjust their senior water rights use to current economic conditions. Water use transfers are an important means for achieving the highest and best use of water. Tribes, however, should not ignore the needs of third parties in tribal water transfers. Notably, as tribal water codes are being drafted, they need to consider these and other third party impacts.

Tribal consideration of transfers of water to off-reservation uses raises additional questions. Foremost among these questions is what law applies to the tribal water when it leaves the reservation. States have very strict rules to protect existing and even junior water uses from injury as a result of a senior water right transfer. State law, however, does not apply to Indians' use of their water on their reservations. Federal and tribal laws governing Indian tribes' use of their water are still evolving but are beginning to conflict with state attempts to extend its water laws into Indian country. This fact, along with the specter of non-Indians having to compete and pay for water they used historically for free, is causing state and tribal tensions.

Indian reservations suffer the highest rates of poverty and attendant social ills of any region in this country. Tribal governments therefore must find ways to develop healthy economies and employment. With declining federal funding, tribal water marketing can generate critically needed capital for development. Leasing opportunities exist both on and off reservations, and several recent settlements include provisions for off-reservation leasing of tribal water.

RECOMMENDATIONS:

- When Indian water rights are transferred or applied to off-reservation uses, tribal governments should establish procedures to evaluate third party effects.
- State and federal governments should cooperate with tribes in evaluating and implementing mutually beneficial transfers.

Conclusion 9: Water laws should be enacted to promote water conservation and salvage while protecting third party interests.

Water is a scarce resource in the West, and all intentional and nonintentional uses of water that prohibit reuse merit careful attention. Interest in water conservation and salvage operations has been aroused by the Imperial Irrigation District to Metropolitan Water District transfer described in Chapter 11. Although it may be difficult to duplicate the success of this transfer, countless smaller salvage projects could be undertaken. In addition, some states—particularly Arizona—are looking to agricultural water conservation to make water available for other uses and have enacted policies to encourage or mandate agricultural conservation.

Technology is available to reduce much water waste. Because much of the water in the West is used by irrigated agriculture, however, the feasibility of applying the technology is limited by the economic condition of agriculture. The feasibility of applying technology to reduce evaporation losses from water surfaces, reduce transpiration from unwanted plants, and reduce loss of water to areas where it is not recoverable is limited by economics. Salvage is further complicated by legal uncertainties concerning who owns or controls the conserved water; state laws that prohibit the transfer, sale, or reuse of conserved water; and the difficulties of maintaining the timing and quantity of return flows on which downstream rights holders depend.

If widespread water conservation and salvage are to take place throughout the West, the return flow problem must first be overcome. The principal reason the Imperial Irrigation District/Metropolitan Water District salvage project succeeded is because there are no downstream senior water rights that depend on the return flows, with the exception of environmental third parties (e.g., the Salton Sea).

If legal disincentives to water conservation and salvage are removed and unnecessary procedural requirements are relaxed, the cost of converting wasted water to productive uses could be reduced. Some controls are necessary, of course, such as existing laws that prevent a would-be salvager from depriving another water user of a water supply by consuming more water than the amount historically used or allowed by an existing water right. But states should act to remove the threat that salvaged water will be declared abandoned. They may want to liberalize the standard of proof that water is truly conserved or salvaged and not being taken from others. Water is, after all, the property of the public, and a public service is provided by those whose conservation and salvage efforts make more water available in a stream.

The great potential of water salvage notwithstanding, laws that facilitate salvage should not ignore the potential for disruption of natural systems, environmental amenities, and the expectations of future water users. Improvements in irrigation efficiency can dry up a wetland. Streams valuable for river rafting can be reduced to unsuitable levels. These interests should be weighed in an established state transfer process that includes procedures for considering effects on the public interest. To the extent that these matters are generically addressed in the allocation and transfer process, no special treatment is necessary for a transfer or reuse of conserved or salvaged water. A state, however, may wish to use the transfer of conserved or salvaged water as an opportunity to further public purposes. For instance, Oregon has integrated the goals of its instream flow protection program with a conservation and salvage law enacted in the late 1980s. It requires that 25 percent of the water conserved be allocated to the state to maintain instream flow. Thus far, however, the committee is not aware of any transfers that have occurred under Oregon's new law.

RECOMMENDATIONS:

- States, tribal governments, and federal agencies should establish programs to reduce the uncertainties and costs involved in water conservation and salvage, facilitating and providing incentives to encourage conservation and reuse of water.
- State, tribal, and federal water transfer review processes should take full account of the third party effects of a transfer of conserved or salvaged water, just as with any other transfer.

Conclusion 10: Water transfer reviews should consider the interrelationships between water quality and water quantity and also between surface and ground water resources.

Where water is scarce, it is particularly important to recognize the relationships between quantity and quality and between surface water and ground water supplies. As population growth and economic development continue in the West, the increased demand for water is likely to highlight these relationships. For instance, both sewage effluent and irrigation return flows are potential causes of water quality degradation, depending on the degree to which they are diluted with higher-quality water. Uncertainty about the water rights status of sewage effluent in some states (e.g., Arizona) restricts water use changes that would allow use of effluent for replacement purposes or to meet downstream senior water rights.

Growth in the urban and industrial sectors also can bring increased pollution, and polluted supplies in effect reduce the quantity of water available for use. Effects on ground water and surface water quality may arise from transfers to new locations and uses, and from the diminished assimilative capacity of the water bodies from which the transfers are made.

RECOMMENDATIONS:

- States and tribes should accelerate the development of laws, policies, and administrative procedures to ensure consideration of water quality changes that might result from transfers involving the use of municipal effluent and other waters of impaired quality.
- States and tribes should encourage conjunctive management of surface and ground water, based on knowledge of hydrologic connections and a full consideration of the related water quality issues and other potential third party effects.
- States and tribes should require agencies to develop technical capabilities for evaluating and monitoring surface and ground water quality as part of the transfer evaluation process.

Conclusion 11: Federal legislative and administrative policies should more clearly support federal water transfers while addressing third party effects and the distribution of benefits from transfers involving federal project water.

Current policies of the Congress and the U.S. Department of the Interior regarding the transfer of entitlements to federal project waters are vague. Although they represent a positive step to separate project entitlements from the original beneficiaries, the policies provide no guidance on how to identify and evaluate third party impacts when entitlements to use federal project water are transferred. Attitudes toward transfers vary considerably within federal water agencies; some regional offices promote transfers aggressively, whereas others generally take a hands-off attitude or actually restrict transfers. Neither Congress nor federal project administrators has been clear as to which third party effects should warrant denial of federal project transfers and which need to be mitigated before transfers are allowed to proceed.

For example, Congress has not spoken on the crucial issue of the resale of federal waters at a profit—the "windfall" issue. Congress could decide to recapture some of the federal (i.e., taxpayer) investment in federal projects, either in the form of financial contributions or in the form of water that might be reallocated to new purposes such as environmental protection or mitigation.

These uncertainties are an impediment to changes in the use and location of use of federal project water and provide sharp contrast to the orderly review processes that exist in some states (e.g., Idaho).

These uncertainties can lead to inefficient and sometimes environmentally and socially damaging solutions to western water problems.

Federal policies on voluntary water transfers generally defer to state law and administrative practice in determining how the third party effects of proposed water transfers are to be assessed and accommodated. Nevertheless, some federal project consistency should be encouraged in this critical area of concern. Some third party effects are so serious that they may warrant federal denial of proposed water transfers. Others may not warrant denial but may require that substantial mitigation measures be taken.

RECOMMENDATIONS:

- The Secretary of the Interior should develop a formal process for assessing transfers of federal project water. This process should identify the role of those agencies of the Department of the Interior responsible for natural resource stewardship, and allow for consultation with tribal governments, states, and the interested public, to ensure that major third party effects are assessed and mitigated.
- Congress should set clear policy for the distribution of profits from the resale of federal project water, and this policy should provide a portion of the economic value in federal project water to be recaptured for public use.
- The Secretary of the Interior should require greater consistency in applying the department's general policy supporting voluntary transfers and should specify that there is authority to transfer water among as well as within project service areas. Careful attention must be paid to third party effects.

Appendix A

Glossary

ACEQUIA—Community-run irrigation ditches and/or the community-run organizations that manage them. This system of water management is rooted in ancient Spanish custom; many still operate in northern New Mexico.

ACRE-FOOT—The volume of water required to cover 1 acre of land to a depth of 1 foot. Equal to 1.2 megaliters (ML) or 1,233.5 cubic meters (m^3).

APPROPRIATION DOCTRINE—The system of water law dominant in the western United States under which (1) the right to water was acquired by diverting water and applying it to a beneficial use and (2) a right to water acquired earlier in time is superior to a similar right acquired later in time. Also called prior appropriation doctrine. In most states, rights are not now acquired by diverting water and applying it to a beneficial use. Such a system is referred to as the constitutional method of appropriation. Rights are acquired by application, permit, and license, which may not require diversion and application to a beneficial use. Superiority of right is based on earliest in time and has no reference to whether two rights are for a similar use.

AQUIFER—Water-bearing strata of permeable rock, gravel, or sand.

ARID CLIMATE—Generally any extremely dry climate.

BASIN—(1) *Hydrology*: The area drained by a river and its tributaries. (2) *Irrigation*: A level plot or field, surrounded by dikes, which may be flood irrigated. (3) *Erosion control*: A catchment constructed to contain and slow runoff to permit the settling and collection of soil materials transported by overland and rill runoff flows.

BENEFICIAL USE—A use of water resulting in appreciable gain or benefit to the user, consistent with state law, which varies from one state to another.

BENEFIT-COST RATIO—An economic indicator of the efficiency of a proposed project, computed by dividing benefits by costs; usually, both the benefits and the costs are discounted, so that the ratio reflects efficiency in terms of the present value of future benefits and costs.

CANAL—A constructed open channel for transporting water from the source of supply to the point of distribution.

CHANNEL—A natural stream that conveys water; a ditch or channel excavated for the flow of water.

COMMERCIAL WITHDRAWALS—Water withdrawn for use by motels, hotels, restaurants, office buildings, commercial facilities, and civilian or military institutions. The water may be obtained from a public supply, or it may be self-supplied.

CONJUNCTIVE WATER USE—The joining together of two sources of water, such as ground water and surface water, to serve a particular use.

CONSUMPTIVE USE—Use of water that renders it no longer available because it has been evaporated, transpired by plants, incorporated into products or crops, consumed by people or livestock, or otherwise removed from water supplies.

CONVEYANCE LOSS (WATER)—Loss of water from delivery systems during conveyance, including operational losses and losses due to seepage, evaporation, and transpiration by plants growing in or near the channel.

CORRELATIVE RIGHTS—Another term for the reasonable use doctrine relating to percolating and riparian waters. In the ground water context, the doctrine of correlative rights will generally limit the appropriation of ground water to the landowner's proportionate share thereof.

DEPLETION—Net rate of water use from a stream or ground water aquifer for beneficial and nonbeneficial uses. For irrigation or municipal uses, the depletion is the headgate or wellhead diversion minus return flow to the same stream or ground water aquifer.

DISCHARGE (HYDRAULICS)—Rate of flow, especially fluid flow; the volume of fluid passing a point per unit time, commonly expressed as cubic feet per second, million gallons per day, gallons per minute, or cubic meters per second.

DISSOLVED SOLIDS—Minerals and organic matter dissolved in water.

DIVERSION—A turning aside or alteration of the natural course of a flow of water, normally considered physically to leave the natural channel. In some states, this may be a consumptive use direct from a stream, such as by livestock watering. In other states, a diversion must consist of such actions as taking water through a canal, pipe, or conduit.

DOMESTIC WATER USE—Water used for normal household purposes, such as drinking, food preparation, bathing, washing clothes and dishes, flushing toilets, and watering lawns and gardens. Also called residential water use or domestic withdrawals. The water may be obtained from a public supply or may be self-supplied.

DOMESTIC WITHDRAWALS—Water used for normal household purposes, such as drinking, food preparation, bathing, washing clothes and dishes, flushing toilets, and watering lawns and gardens. Also called residential water use or domestic water use. The water may be obtained from a public supply or may be self-supplied.

DRAINAGE BASIN—Land area drained by a river.

DUTY OF WATER—The total volume of water per year that may be diverted under a vested water right.

EMINENT DOMAIN—The authority of the federal or state government, or an agency or party authorized by the federal government, to condemn all interest in land for public purposes, after payment of just compensation.

ESTUARY—That portion of a coastal stream influenced by the tide of the body of water into which it flows, for example, a bay or mouth of a river, where the tide meets the river current; an area where fresh and marine waters mix.

EVAPORATION—Process by which water is changed from a liquid to a vapor. Also see evapotranspiration.

EVAPOTRANSPIRATION—The sum of evaporation and transpiration from a unit land area. Also see consumptive use.

FLOOD IRRIGATION—The application of water in which the entire surface of the soil is covered by a sheet of water, called "controlled flooding" when water is impounded or the flow is directed by border dikes, ridges, or ditches.

FLOODPLAIN—Land bordering a stream and subject to flooding.

FLOW—As used in this report, movement of water.

FRESH WATER—Water that contains less than 1,000 milligrams per liter (mg/L) of dissolved solids; generally more than 500 mg/L is undesirable for drinking and many industrial uses.

GROUND WATER—Generally, all subsurface water as distinct from surface water; specifically, the part that is in the saturated zone of a defined aquifer.

GROUND WATER MINING—The withdrawal of ground water through wells, resulting in a lowering of the ground water table at a rate faster than the rate at which the ground water table can be recharged.

HYDROELECTRIC POWER GENERATION WATER USE—The instream use of surface water to drive turbines and generate electric power, a nonwithdrawal use.

HYDROLOGIC CYCLE—The circuit of water movement from the atmosphere to the earth and return to the atmosphere through various stages or processes, such as precipitation, interception, runoff, infiltration, percolation, storage, evaporation, and transpiration.

IMPOUNDMENT—Generally, an artificial collection or storage of water, as a reservoir, pit, dugout, or sump.

INDUSTRIAL WATER USE—Water used by manufacturing plants, such as steel, chemical and allied products, paper and allied products, mining, and petroleum refining. The water may be obtained from a public supply or may be self-supplied.

INDUSTRIAL WITHDRAWALS—Water withdrawn for or used for thermoelectric power (electric utility generation) and other industrial and manufacturing uses such as steel, chemical and allied products, paper and allied products, mining, and petroleum refining. The water may be obtained from a public supply or may be self-supplied.

INSTREAM FLOW REQUIREMENT—Flow required in a stream to maintain desired instream benefits such as navigation, water quality, fish propagation, and recreation.

INSTREAM FLOW RIGHTS—A doctrine used to preserve minimum river or streamflows for fish and wildlife, recreation, water quality, and scenic beauty, among other public purposes. Such rights are limited to the use of water within its natural course, not requiring diversion.

INSTREAM USE—Any use of water that does not require diversion or withdrawal from the natural watercourse, including in-place uses such as navigation and recreation.

INTERBASIN TRANSFER OF WATER—A transfer of water rights and/or water from the basin of origin to a different hydrologic basin.

INTERMITTENT STREAM—A stream or part of a stream that flows only in direct response to precipitation. It receives little or no water from springs and melting snow, or other sources. It is dry for a large part of the year, generally more than 3 months.

INTERSTATE WATERS—Waters legally defined as rivers, lakes, and other waters that flow across or form a part of state or international boundaries; and coastal waters—whose scope has been defined to include ocean water along the coast—that are influenced by tides.

IRRIGATION—The application of water to soil for crop production or for turf, shrubbery, or wildlife food and habitat. Provides water requirements of plants not satisfied by rainfall.

IRRIGATION DISTRICT—In the United States, a cooperative, self-governing public corporation set up as a subdivision of the state, with definite geographic boundaries, organized to obtain and distribute water for irrigation of lands within the district; created under authority of the state legislature with the consent of a designated fraction of the landowners or citizens and having taxing power.

IRRIGATION EFFICIENCY—Ratio of irrigation water used in evapotranspiration to the water applied or delivered to a field or farm. This is one of several indices used to compare irrigation systems and to evaluate practices.

IRRIGATION FREQUENCY—Time interval between irrigations.

IRRIGATION RETURN FLOW—The part of applied water that is not consumed by evapotranspiration and that migrates to an aquifer or surface water body. See also return flow.

IRRIGATION WATER REQUIREMENT—The quantity, or depth, of water in addition to precipitation, required to obtain desired crop yield and to maintain a salt balance in the crop root zone.

IRRIGATION WITHDRAWALS—Withdrawal of water for application on land to assist in the growing of crops and pastures or to maintain recreational lands.

JUNIOR RIGHTS—A junior water rights holder is one who holds rights that are temporarily more recent than senior rights holders. All water rights are defined in relation to other users, and a water rights holder only acquires the right to use a specific quantity of water under specified conditions. Thus, when limited water is available, junior rights are not met until all senior rights have been satisfied.

MEGALITER—A measure of volume: 1 ML equals 0.8107 acre-foot; 1 acre-foot equals 1.23 ML and is the volume of water required to cover 1 acre of land to a depth of 1 foot.

NET DEPLETION—Total water consumed by irrigation, or other use in an area, equal to water withdrawn minus return flow.

OFF-SITE USE—A use of water away from the point of diversion or withdrawal.

OFF-STREAM USE—Water withdrawn or diverted from a ground or surface water source of use.

OUTLET—Point of water disposal from a stream, river, lake, tidewater, or artificial drain.

PER CAPITA USE—The average amount of water used per person per day.

PERFECTED WATER RIGHT—A water right to which the owner has applied for and obtained a permit, has complied with the conditions of the permit, and has obtained a license or certificate of appropriation.

PERMIT SYSTEM—A general term referring to a system of acquiring water rights under state law whereby the state must issue a permit for a new use of water; although permit systems were, at one time, generally associated with eastern states using the riparian doctrine, they are now found elsewhere as well.

PLACE-OF-USE LIMITATION—In the water law context, the act of defining a water right so that the owner of the right may not freely change the place of use without consideration of the effect of such change on other water users.

POINT SOURCE OF POLLUTION—Pollution originating from any discrete source, such as the outflow from a pipe, ditch, tunnel, well, concentrated animal-feeding operation, or floating craft.

PRECIPITATION—Includes rain, snow, hail, and sleet.

PRIOR APPROPRIATION—A concept in water law under which a right is determined by such a procedure as having the earliest priority date.

PRIORITY OF USE AND STATUTORY PREFERENCES—Under appropriation water law systems, priority of use refers to the date a water right is acquired, with senior rights prevailing over junior rights. Priority of use must be distinguished from statutory preferences, which refer to statutory statements of preference among types of beneficial use and would come into play, for example, in deciding which of two concurrent water rights should be satisfied first during a shortage of water or which of two competing applications for a water right should be granted.

PUBLIC INTEREST—An interest or benefit accruing to society generally, rather than to any individuals or groups of individuals in the society.

PUBLIC SUPPLY—Water withdrawn for all uses by public and private water suppliers and delivered to users that do not supply their own water. Water suppliers provide water for a variety of uses such as domestic, commercial, industrial, and public water use.

PUBLIC TRUST DOCTRINE—A poorly defined judicial doctrine under which the state holds its navigable waters and underlying beds in trust for the public and is required or authorized to protect the public interest in such waters. All water rights issued by the state are subject to the overriding interest of the public and the exercise of the public trust by state administrative agencies.

PUBLIC WATER USE—Water from a public supply used for fire fighting, street washing, and municipal parks and swimming pools. See also public supply.

RAINFALL—Quantity of water that falls as rain only. Not synonymous with precipitation.

REACH—A specified length of a stream or channel.

REASONABLE USE—A rule with regard to percolating or riparian water restricting the landowner to a reasonable use of his own rights and property in view of and qualified by the similar rights of others, and the condition that such use not injure others in the enjoyment of their rights.

RECHARGE—Process by which water is added to the zone of saturation, as recharge of an aquifer.

RECHARGE AREA (GROUND WATER)—An area in which water infiltrates the ground and reaches the zone of saturation.

RECLAIMED SEWAGE—Treatment plant effluent that has been diverted or intercepted for use before it reaches a natural waterway or aquifer.

RECYCLED WATER—Water that is used more than one time by the same user.

REGULATION OF A STREAM—Artificial manipulation of the flow of a stream.

RENEWABLE WATER SUPPLY—The rate of supply of water (volume per unit time) potentially or theoretically available for use in a region on an essentially permanent basis.

RESERVED WATER RIGHTS—This class of water rights is a judicial creation derived from *Winters* v. *United States* (207 U.S. 564, 1907) and subsequent federal case law, which collectively hold that when the federal government withdraws land from general use and reserves it for a specific purpose, the federal government by implication reserves the minimum amount of water unappropriated at the time the land was withdrawn or reserved to accomplish the primary purpose of the reservation. Federal reserved water rights may be claimed when Congress has by statute withdrawn lands from the public domain for a particular federal purpose or where the President has withdrawn lands from the public domain for a particular federal purpose pursuant to congressional authorization.

RETURN FLOW—The amount of water that reaches a ground or surface water source after release from the point of use and thus becomes available for further use. See also irrigation return flow.

REUSE—Repeated use of the same water by subsequent users in sequential systems.

RIPARIAN DOCTRINE—The system of law dominant in Great Britain and the eastern United States, in which owners of lands along the banks of a stream or water body have the right to reasonable use of the waters and a correlative right protecting against unreasonable use by others that substantially diminishes the quantity or quality of water. The right is appurtenant to the land and does not depend on prior use.

RIPARIAN LAND—Land situated along the bank of a stream or other body of water.

RIPARIAN RIGHTS—A concept of water law under which authorization to use water in a stream is based on ownership of the land adjacent to the stream and normally not lost if not used.

RUNOFF—That part of the precipitation that moves from the land to surface water bodies.

SAFE YIELD (GROUND WATER)—Amount of water that can be withdrawn from an aquifer without producing an undesired effect.

SAFE YIELD (SURFACE WATER)—Amount of water that can be withdrawn or released from a reservoir on an ongoing basis with an acceptably small risk of supply interruption (reducing the reservoir storage to zero).

SALINITY—Concentration of dissolved salts in water or soil water.

SALVAGE— Saving or conserving water by improving the efficiency of use (e.g., by lining irrigation canals so seepage losses are reduced).

SENIOR RIGHTS—A senior water rights holder is one who holds rights that are older than those of junior rights holders. All water rights are defined in relation to other users, and a water rights holder only acquires the right to use a specific quantity of water under specified conditions. Thus, when limited water is available, senior rights are satisfied first.

STREAM DEPLETION—See depletion.

SURFACE WATER—An open body of water such as a stream, lake, or reservoir.

THIRD PARTIES—The people, communities, and environments not directly engaged in a transfer of water or water rights (i.e., the buyers or sellers) but still affected by the transfer. These affected parties can include areas of origin, Indian tribes, other minority cultures (primarily Hispanic), communities that depend on irrigated agriculture or water-based recreation, boaters, anglers, and broad segments of the public who care about wetlands, riparian areas, endangered species, instream flows, and other environmental values that might be harmed or enhanced by a change in water use.

TRANSFER—A movement of water or water rights that involves a change in point of diversion, a change in type of use, or a change in location of use.

TRANSFER (CONVEYANCE OF WATER RIGHT)—A passing or conveyance of title to a water right; a permanent assignment as opposed to a temporary lease or disposal of water.

WASTEWATER—Water that carries wastes from homes, businesses, and industries; a mixture of water and dissolved or suspended solids.

WATER CONSUMED—See consumptive use.

WATER DEMAND—Water requirements for a particular purpose, such as irrigation, power, municipal supply, plant transpiration, or storage.

WATER MANAGEMENT—Application of practices to obtain added benefits from precipitation, water, or water flow in any of a number of areas, such as irrigation, drainage, wildlife and recreation, water supply, watershed management, and water storage in soil for crop production. Includes irrigation water management and watershed management.

WATER QUALITY—The chemical, physical, and biological condition of water related to beneficial use.

WATER RESOURCE—The supply of ground and surface water in a given area.

WATER RESOURCE REGION—Natural drainage basin or hydrologic area that contains either the drainage area of a major river or the combined areas of a series of rivers.

WATER RESOURCE SUBREGION—A subregion of a water resource region that includes the area drained by a river system, a reach of a river and its tributaries in that reach, a closed basin(s), or a group of streams forming a coastal drainage area.

WATER RIGHT—The legal right to use a specific quantity of water, on a specific time schedule, at a specific place, and for a specific purpose.

WATER RIGHTS, CORRELATIVE DOCTRINE—When a source of water does not provide enough for all users, the water is reapportioned proportionately on the basis of prior water rights held by each user.

WATER USE EFFICIENCY—Marketable crop production per unit of water consumed in evapotranspiration.

WATER WITHDRAWAL—Water removed from ground or surface water for use.

Appendix B

Acknowledgments of Case Study Participants

The committee members wish to extend their sincere appreciation to all the people who shared their expertise and experience with us during the preparation of this report. In particular, we would like to thank Joseph Sax, a leading water law scholar from the University of California, Berkeley, for his important insights and contributions. We also would like to acknowledge the representatives who met informally with us as we gathered information for the case studies—people representing agricultural interests, water management agencies, urban planners, environmental groups, Indian tribes, and other interests as appropriate. Our discussions were frank and informative. In the end, however, the committee claims sole responsibility for the content and recommendations of this report.

TRUCKEE-CARSON BASINS IN NEVADA

RON ANGLIN, Stillwater Wildlife Refuge, Fallon, Nevada
CHESTER BUCHANAN, U.S. Fish and Wildlife Service, Reno, Nevada
FRANK DIMICK, U.S. Bureau of Reclamation, Carson City, Nevada
JOSEPH ELY, Pyramid Lake Paiute Tribe, Nixon, Nevada
ROBERT FIRTH, Westpack Utilities, Reno, Nevada
LYMAN MCCONNELL, Truckee-Carson Irrigation District, Fallon, Nevada
PETE MORROS, Nevada State Engineer, Carson City, Nevada

GARY STONE, Federal Water Master, Reno, Nevada
DAVID YARDAS, Environmental Defense Fund, Oakland, California

COLORADO FRONT RANGE AND ARKANSAS RIVER VALLEY

JOHN CARLSON, Carlson, Hammond, Paddock, Denver, Colorado
JERIS DANIELSON, Colorado Department of Natural Resources, Denver, Colorado
MAX DOTSON, U.S. Environmental Protection Agency, Denver, Colorado
ROLLIE FISCHER, Colorado River Water Conservation District, Glenwood Springs, Colorado
TOM GRISWOLD, City of Aurora Public Utilities Department, Aurora, Colorado
GREG HOBBS, Davis, Graham and Stubbs, Denver, Colorado
DANIEL LUECKE, Environmental Defense Fund, Boulder, Colorado
BILL MILLER, Denver Water Board, Denver, Colorado
DAVE SHELTON, Colorado Department of Health, Denver, Colorado
LARRY SIMPSON, Northern Colorado Water Conservancy District, Loveland, Colorado
TOMMY THOMSON, Southeastern Colorado Water Conservancy District, Pueblo, Colorado

NORTHERN NEW MEXICO

LILA BIRD, Water Information Network, Albuquerque, New Mexico
WALTER DASHENO, Eight Northern Pueblos Indian Council, San Juan Pueblo, New Mexico
ELUID MARTINEZ, State Engineers Office, Santa Fe, New Mexico
WILFRED RAEL, Questa, New Mexico
RICHARD ROSENSTOCK, Santa Fe, New Mexico
FRED WALTZ, Taos, New Mexico
PETER WHITE, Santa Fe, New Mexico

THE YAKIMA BASIN IN WASHINGTON

URBAN EBERHART, Yakima River Basin Association of Irrigation Districts, Allensburg, Washington
LEVI GEORGE, Yakima Indian Nation, Toppenish, Washington
JOHN KEYS, Bureau of Reclamation, Boise, Idaho

KAHLER MARTINSON, Department of Fisheries, Olympia, Washington
DAVID ORTMAN, Friends of the Earth, Seattle, Washington
KENNETH SLATTERY, Department of Ecology, Olympia, Washington

CENTRAL ARIZONA

JONI BOSH, Sierra Club, Phoenix, Arizona
BARTLEY P. CARDON, Tucson, Arizona
HERB DISHLIP, Department of Water Resources, Phoenix, Arizona
HERB GUENTHER, Arizona State Legislature, Phoenix, Arizona
HELEN INGRAM, University of Arizona, Tucson, Arizona
JOHN LESHY, Arizona State University, Tempe, Arizona
ROD LEWIS, Gila River Indian Community, Sacaton, Arizona
ROGER MANNING, Arizona Municipal Water Users Association, Phoenix, Arizona
AUSTIN NUNEZ, Tohono O'Odham Nation, Tucson, Arizona
JACKIE RICH, Rich and Associates, Phoenix, Arizona
BETSY RIEKE, Jennings, Strouss, Salmon, Phoenix, Arizona
JOHN SCHMAUCHER, Colorado River Indian Tribes, Parker, Arizona

CALIFORNIA'S CENTRAL VALLEY

HAMILTON CANDEE, Natural Resources Defense Council, San Francisco, California
DAVID CONE, Broadview Water District, Firebaugh, California
MARTHA DAVIS, Mono Lake Committee, Los Angeles, California
BRIAN GRAY, Hastings College of Law, San Francisco, California
STEVEN HALL, California Farm/Water Coalition, Fresno, California
LARRY HANCOCK, Bureau of Reclamation, Sacramento, California
JERRY JOHNS, State Water Resources Control Board, Sacramento, California
RON KHACHIGIAN, Blackwell Land Company, Bakersfield, California
MITCH KODAMA, Los Angeles Department of Water and Power, Los Angeles, California
DICK MOSS, Friant Water Users Association, Lindsay, California
BOB POTTER, Department of Water Resources, Sacramento, California
STUART PYLE, Kern County Water Agency, Bakersfield, California

ANNE THOMAS, Best, Best and Krieger, Riverside, California
DON VILLAREJO, California Institution for Rural Studies, Davis, California

CALIFORNIA'S IMPERIAL IRRIGATION DISTRICT AND METROPOLITAN WATER DISTRICT

JERALD DAVISSON, Palo Verde Irrigation District, Blythe, California
KIRK DIMMICK, Metropolitan Water District of Southern California, Los Angeles, California
ED HELENBECK, Bureau of Reclamation, Boulder City, Nevada
TOM LEVY, Coachella Valley Water District, Coachella, California
JAN MATUSAK, Metropolitan Water District of Southern California, Los Angeles, California
CHARLES SHREVE, Imperial Irrigation District, Imperial, California
RANDAL STOCKER, Imperial Irrigation District, Imperial, California
FRED WORTHLEY, Department of Fish and Game, Long Beach, California

Biographical Sketches of Committee Members and Staff

A. DAN TARLOCK, *Chair,* is a professor of law at Illinois Institute of Technology Chicago-Kent College of Law. He received his LL.B. from Stanford University. His professional experience includes private practice, consulting, and teaching. Mr. Tarlock has authored and coauthored many publications concerning water resource management, land use, and environmental law and policy, including a treatise and a casebook on water law. He is the vice-chair of the Water Science and Technology Board (WSTB) and a member of its Committee to Review the Glen Canyon Environmental Studies.

D. CRAIG BELL, executive director, Western States Water Council, received his law degree from the University of Utah in 1973. He has served as secretary-treasurer and member of the Board of Trustees of Water & Man, Inc. He has published numerous articles on water rights, Indian water rights, and the states' role in national water policy.

BONNIE COLBY, associate professor of agricultural economics at the University of Arizona (1983 to present), received her Ph.D. in agricultural and resource economics from the University of Wisconsin. She previously served as a policy analyst with the Department of Food and Agriculture, state of California. Her research and teaching focus on the economics of natural resource management and on valuation of environment quality, natural amenities, and water rights. Dr. Colby has been one of the leading scholars in the area of water reallocation and was major author of *Water Markets in Theory and Practice*, a book published in 1987. She is a member of the WSTB's Committee to Review the Glen Canyon Environmental Studies.

LEO M. EISEL, of Wright Water Engineering in Denver, Colorado, received his Ph.D. in engineering from Harvard University in 1970. From 1971 to 1973 he was a staff scientist with the Environmental Defense Fund in New York; later he became director of the Illinois Division of Water Resources, and from 1977 to 1980 was director of the U.S. Water Resources Council. He is a former WSTB member and was a member of the Committee to Review the Metropolitan Washington Area Water Supply Study. Dr. Eisel is broadly experienced in water supply planning and hydrologic engineering.

DAVID H. GETCHES, professor of law, University of Colorado School of Law (1979 to present), received his J.D. from the University of Southern California. Previous positions include executive director, State of Colorado Department of Natural Resources, 1983 to 1987; founding director, Native American Rights Fund, Boulder, Colorado, 1970 to 1976; directing attorney, California Indian Legal Services, 1968 to 1970; and private practice. Mr. Getches has an extensive publication list, especially in the water resource management and Indian law areas.

THOMAS J. GRAFF, attorney, Environmental Defense Fund, received his law degree from Harvard Law School in 1967. Previous positions include lobbyist for New York City and clerk for a federal judge and teaching law at the University of California, Berkeley, and Harvard. He was a governor's appointee to the Colorado River Board of California and a member of the Citizens Advisory Committee to the San Joaquin Valley Drainage Program.

FRANK GREGG, professor, Department of Renewable Natural Resources, University of Arizona, is a former director of the Bureau of Land Management and former chairman of the New England River Basins Commission. He recently directed the first phase of a continuing study of severe drought coping strategies for the U.S. Southwest and is principal investigator of a study now nearing completion of institutional innovation in the field of water resources. He served on the Policy Subcommittee of the National Academy of Sciences' Committee on Irrigation-Induced Water Quality Problems.

R. KEITH HIGGINSON, director, Idaho Department of Water Resources, received a B.S. in civil engineering from Utah State University in 1957. His professional career has included water resource consulting engineering; water and agricultural research; commissioner, Bureau of Reclamation (1977 to 1981); and more than 23 years as a state water resource administrator. He is a former WSTB member and has also been a member of a number of interstate water resource boards and commissions.

MARVIN E. JENSEN, director, Colorado Institute for Irrigation Management, Colorado State University, has a Ph.D. in civil engineering from Colorado State University. He was previously national program leader, Water Management/Salinity Program (1979 to 1987), and director, Snake River Conservation Research Center (1969 to 1978). He served as chairman of the Council for Agriculture Science and Technology Task Force on Effective Use of Water in Irrigated Agriculture (1988) and as senior editor of *ASCE Manual on Evapotranspiration and Irrigation Water Requirements* (1990). He is a member of the National Academy of Engineering.

DUNCAN T. PATTEN, professor of botany and director of the Center for Environmental Studies, Arizona State University (ASU), received his Ph.D. from Duke University. Other academic experience includes assistant professor of botany at Virginia Polytechnic Institute and associate professor and assistant academic vice president at ASU. He is senior scientist for the U.S. Department of Interior's Bureau of Reclamation Glen Canyon research program. He served on the Board on Environmental Studies and Toxicology (BEST) and the WSTB Committee to Review Glen Canyon Environmental Studies. He chaired the BEST committee on the Mono Basin Ecosystem and is on BEST's committee to study science in the national parks. He is also a member of the National Research Council's Commission on Geoscience, Environment, and Resources.

CLAIR B. STALNAKER, chief, Riverine and Wetland Ecosystems Branch, National Ecology Research Center, U.S. Fish and Wildlife Service; adjunct professor, Department of Fisheries and Wildlife, Colorado State University; and adjunct professor, Department of Civil Engineering, Utah State University, received a B.S. in forestry from West Virginia University and a Ph.D. in animal ecology from North Carolina State University in 1966. Previous experience with the U.S. Fish and Wildlife Service includes leader, Utah Cooperative Fishery Research Unit; leader, Instream Flow and Aquatic Systems Group;

fishery research specialist, Division of Federal Assistance and Endangered Species; and fishery research biologist, Western Water Allocation Program.

LUIS S. TORRES is a staff member with Southwest Research and Information Center, Inc. He is currently working under a Ford Foundation grant in northern New Mexico, providing water management education and technical assistance to acequias and other community groups. He has worked continuously in community-assistance-related work in northern New Mexico for the past 18 years, including 10 years as program director of American Friends Service Committee.

RICHARD TRUDELL, a member of the Santee Sioux Tribe in Nebraska, is executive director and principal founder of the American Indian Lawyer Training Program, Inc. (AILTP). Mr. Trudell has served on the boards of numerous organizations, including the Executive Committee and Board of Trustees of the Robert F. Kennedy Memorial and the National Board of Directors of the Legal Services Corporation under an appointment by President Carter. He is a member of the American and Nebraska bar associations and has been active in organized bar activities. He was appointed by the president of the American Bar Association to serve on its Commission on Opportunities for Minorities in the Profession. He received his law degree from Catholic University in 1972.

HENRY J. VAUX, JR., professor of resource economics, Department of Soil and Environmental Sciences, University of California, Riverside, received a Ph.D. in natural resource economics from the University of Michigan in 1973. He has been director of the Water Resources Center since 1986 and has been employed by the University of California since 1970. He was an economist with the National Water Commission and an examiner in the Resources and Civil Works Division of the Office of Management and Budget. Dr. Vaux has written extensively on the economics of land and water resources.

SUSAN WILLIAMS, attorney, Gover, Stetson, Williams & West, received her law degree from Harvard Law School in 1981. Previous experience includes lecturer in law, Harvard Law School; adjunct professor of law, Arizona State University; and chair and executive director of the Navajo Tax Commission. She has extensive experience dealing with Indian water rights and resources.

CHRIS ELFRING, a senior program officer at the Water Science and Technology Board, served as study director for the Committee on Western Water Management. She has worked with the National Research Council in various capacities since 1987, and has directed studies of irrigation-induced water quality problems, soil and water research priorities for developing countries, climate change and water resource management, and ground water recharge.

ANITA HALL is an administrative secretary for the Water Science and Technology Board and served as project assistant to the Committee on Western Water Management. She has been on the staff of the National Research Council since 1987.

Index